A BRIEF GUIDE TO SOURCES OF
METALS INFORMATION

A BRIEF GUIDE TO SOURCES OF

METALS

INFORMATION

by

Marjorie R. Hyslop

INFORMATION RESOURCES PRESS
WASHINGTON, D.C.
1973

Available from
Information Resources Press
2100 M Street, N.W.
Washington, D.C. 20037

Library of Congress Catalog Card Number 72-87893

ISBN 0-87815-008-0

CONTENTS

PREFACE

"N ever Give a Searcher an Even Break" is rule 9 of Robert L. Birch's "Bibliographic Suicide-Guide."[1] He quotes McGurk's Law that "Whatever would maximally foul things up is maximally likely to happen." A Briton, Miss B. Smith, latched on to this and, in 1967, applied McGurk's Law to the metallurgical literature[2] (see Figure 1). Since that time, the Institute of Metals (London) has added *Metal Science Journal* to its roster of primary publications, and *Metallurgical Abstracts* has merged with *Review of Metal Literature* (often designated simply as *Metals Review*) to form *Metals Abstracts*. *Metallurgical Abstracts* has been abbreviated during two of its phases as *Met.Abs.* and *M.A.*, respectively, and *Metals Abstracts* is now abbreviated as *Met.A.* Small wonder that the metallurgical engineer is confused.

This book is addressed primarily to the metallurgist and may seem elementary to many librarians and information specialists. However, even the professional librarian should find it useful when faced with a search in some way concerned with metals.

"Metallurgical interest," as treated in this book, encompasses the production, fabrication, treatment, finishing, properties, and applications of metals.[3] Exclusions are mining, production, sta-

[1] Birch, R. L. "Bibliographic Suicide-Guide: For Publishers and Authors," *American Documentation*, 17(1):46, January 1966.

[2] Smith, B. "McGurk's Law and the Metallurgical Literature," *Metallurgia*, 34(6):274, June 1967.

[3] Hyslop, M. R. "What is Metallurgy Today," *Metals Review*, 36(11):7–14, November 1963.

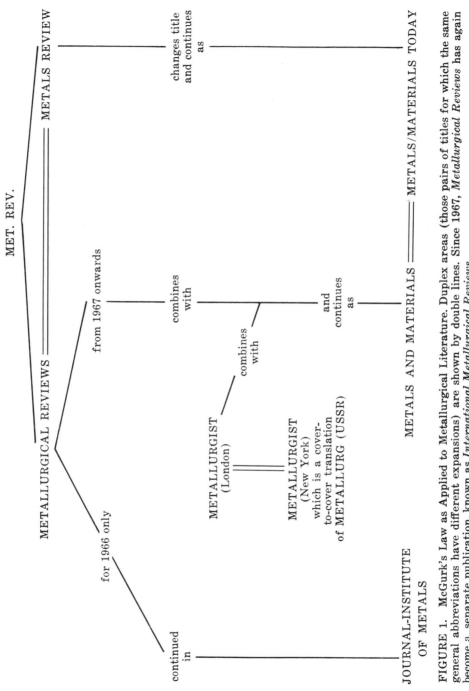

FIGURE 1. McGurk's Law as Applied to Metallurgical Literature. Duplex areas (those pairs of titles for which the same general abbreviations have different expansions) are shown by double lines. Since 1967, *Metallurgical Reviews* has again become a separate publication, known as *International Metallurgical Reviews*.

Reprinted by kind permission of *Metallurgia & Metal Forming*.

tistics, and marketing. While the trade associations listed in Chapter 1 provide production and marketing information, along with technical services, the former is not specifically treated in this book.

Services in which metallurgical interest is peripheral are also excluded. An example is the National Library of Medicine's MEDLARS (Medical Literature Analysis and Retrieval System). Information on toxic effects of various metals could be better searched through this resource than through the more metallurgically oriented services, but it is excluded because a halt had to be called somewhere in covering the overlap of metals with other major fields.

No effort has been made to categorize information sources by individual metals or groups of metals, or by fields of interest. However, the index includes both subjects and titles, so that reference to such terms as copper, aluminum, fatigue, heat treatment, or crystal structure will lead the reader to the principal organizations serving these subject interests in one way or another.

The arrangement of this text basically follows the format established in *A Brief Guide to Sources of Scientific and Technical Information* by Saul Herner (G2), with a few deviations and embellishments. In dealing with a single discipline, as opposed to the whole of science and technology, it has been possible to provide considerable detail about more types of information resources. The greatest deviation from the format of the Herner book is the segregation of all resources in a separate directory section at the end of the book rather than including selective descriptions in each chapter. This has made it possible to concentrate all of the information about a particular organization in a single place and then to refer to the various components in the appropriate chapters. For example, the American Society for Metals appears under "Associations" in the directory, but its library is described in Chapter 1, its abstract journals and bibliographies in Chapter 2, its primary publications in Chapter 3, its searching service in Chapter 5, and self-helps in Chapter 6. This arrangement has its disadvantages, but it is hoped that the nuisance of page-thumbing back and forth from text to directory will be more than offset by the convenience of having all information about a particular organization in one place.

For greater convenience, many of the chapters list pertinent organizations or services by name only and indicate the directory number where full descriptions of each can be found. The directory section lists organizations alphabetically under the six categories, A–F, noted on the contents page. Inevitably, some fall into no precise category, and it is hoped that the catchall of "Other Nonprofit Organizations and Miscellaneous Sources" will not be looked upon as a stigma.

The last directory section, G, is a list of general reference books rather than a list of organizations. Most references cited were used as basic sources of information and can be consulted for additional data or for similar data about other organizations and services not covered in this book. This section provides a number of "guides to metallurgical information" (G10, G11, G13, G15, G21) published in the past, and the reader may wonder "Why another?" One reason is the rise of the information center—a very new phenomenon—and the changing pattern of information resources, particularly those that are computer-based.

The coverage of this book is limited primarily to the United States and Canada. The exceptions consist of several British services—primarily publications—that are widely used in this country. Publications and services in foreign languages are omitted, even though there are some rather important ones in the metallurgical field such as *Bulletin Signalétique*, 740, *Métaux— Métallurgie*, a French abstracting journal that closely parallels the American-British *Metals Abstracts*; and the CEC Metallurgical Documentation and Information System, a pilot scheme for providing computer indexing and searching services to information centers in six European Common Market countries that are currently members of the Commission of the European Communities. An attempt has been made to cover English-language publications and services whatever their source, although U.S. and Canadian resources are given greater in-depth treatment than are those overseas.

I have omitted prices of products and services because, although it is important for the user to know what a product or service sells for, prices are so subject to change—and some price schedules, particularly for products of technical societies, are so complicated—that to show them might mislead rather than inform. As a substitute, pricing is discussed in a general way in each chapter, with enough specificity to give the reader some idea of whether or not a desired service or product is commensurate with his pocketbook.

In general, there is a tendency to think of information services

—particularly library services—as free. But, with the rapid growth of published information in the last three decades, the costly complexities of computer operations, and the general reduction of government support for information activities, charges are now frequently imposed, and the scientific community is becoming educated to expect them.

The directory descriptions show "user qualifications" for some organizations and not for others. Generally, they are explicit for government agencies and information centers. Associations, however, are rather similar in their user restrictions, as covered in Chapter 1. When a library is "open to the public," it should be assumed that the general public refers to individuals with a scientific or engineering background or with a genuine interest in line with their career or occupational activities.

The many organizations that were contacted were most cooperative, and their assistance is deeply appreciated. If some organizations seem to be more generously treated than others, this is not an indication of favoritism, but usually reflects the fact that more complete data were made available by them.

Finally, I am deeply grateful to the American Society for Metals for its cooperation in providing facilities and materials. Its library was indispensable, as was the assistance of its Librarian, Miss Barbara Ladd, who almost invariably came up with a needed resource—or at least knew where to find it.

Chagrin Falls, Ohio
October 1972

1

LIBRARIES AND TECHNICAL SOCIETIES

Oₙₑ metallurgist may find his information needs satisfactorily served by an abstracting journal and an accompanying photoduplication service; another may find that a single information center furnishes all that he wants or can use; still another may rely entirely on a few key journals that cross his desk. Nevertheless, every metallurgist will at some time have an information need that will send him to a library.

A technical society is the basic facility whereby the scientist or technologist nurtures his professional development. A major difference between a library and a technical society is that generally the user must go to the library, whereas the technical society comes to him (with the exception of the meetings, conferences, and seminars which the gregarious American thrives upon and which generate so much of the primary information resource materials). Another difference is that the library tends to be more versatile and broader in its resources, whereas the technical society has a tendency to pinpoint the narrower, or concentrated and detailed, bits and pieces of information.

LIBRARIES

It is probably safe to say that the typical metallurgist knows comparatively little about library resources. Most university curricula, even in engineering schools, require a short course on library usage, but seldom does the student make practical use of this knowledge other than to make a literature search for a master's or doctor's thesis. He usually starts his career lacking full appreciation of the varied library resources that are available to help him in the everyday business of earning a living.

The first admonition in a quest for information is "ask the librarian." Even if you think the possibilities are very remote that what you want will be in the library, or if you're not quite sure what it is you are really looking for—ask anyway. You might be surprised. Librarians have an uncanny ability to find things, or to find out where things are and how to get them. They also have a keen insight that helps them to interpret technical problems and, perhaps by trial and error rather than by full comprehension, to come up with some sort of answer.

For those who want to do some searching on their own, Bloomfield's *How to Use a Library*[1] is helpful. The Crerar Library (A7) describes its collections in the *Guide to Metals Literature,* and *The Book Collection and Services of the Linda Hall Library* (A9) has a useful section on how to use a library card catalog.

If you do not have a library in your own organization, you can try the Science-Technology Division of your public library, the library of a local industrial company or possibly a university in your area, or one of the libraries listed in directory section A.

Libraries serving metallurgical interests may be public, industrial, association (including research institutes), university, or federal. The availability of their services to the general public will vary, but a few generalities can be set forth and some examples given.

Public Libraries: The metallurgical holdings of public libraries will be strongly influenced by local industry. It would be pointless to list all the public libraries in the major American industrial cities where metals are produced, fabricated, or used. A few "special" public libraries with particularly strong metallurgical collections are described in directory section A. They are:

Carnegie Library of Pittsburgh (A5)
Engineering Societies Library (A6)
John Crerar Library (A7)
Library of Congress (A8)
Linda Hall Library (A9)
National Science Library of Canada (A12)

Industrial Libraries: Availability of the services of industrial libraries will depend greatly on the largesse and sense of civic

[1] Bloomfield, M. *How to Use a Library—a Guide for Literature Searching.* Reseda, Calif., Mojave Books, 1970.

duty of the management. They are certainly *not* open to the public. However, if you have a legitimate question that can be quickly answered by phone, or if you can personally look through the library's holdings without bothering the librarian too much, it is worth a try. The major metalworking companies that have good libraries are not listed in this book, but most can be found in the *Directory of Special Libraries and Information Centers* (E7 and G7).

Association Libraries: Most association libraries are operated as a service to the association members—or as repositories of association publications—but policies for their use by the general public are likely to be more liberal than those of industrial libraries. Check the list of associations in directory section B for your subject interest and for user qualifications.

User qualifications for research institutes are likely to be similar to those for associations—usually open to the general public by appointment. A few typical ones with good metallurgy collections are:

Battelle Memorial Institute (F1)
Battelle-Northwest Technical Library (A4)
Ontario Research Foundation Library (A14)
Southern Research Institute Library (A15)
Southwest Research Institute Library (A16)

University Libraries: The university library probably leans just a little bit to the liberal side of the association library. If you have a legitimate need to know, can state your problem precisely (either by phone or in person), and can avoid spending too much of the librarian's time, you are likely to get a cordial reception. The following list of universities in the United States and Canada offering metallurgical curricula (ergo, good library collections) is taken from *Metallurgy/Materials Education Yearbook* published by the American Society for Metals (B15).

University of Alabama
University, Ala. 35486

University of Alberta
Edmonton, Alb., Canada

University of Arizona
Tucson, Ariz. 85721

University of British Columbia
Vancouver 8, B.C., Canada

Polytechnic Institute of Brooklyn
Brooklyn, N.Y. 11201

Brown University
Providence, R.I. 02912

University of California
Berkeley, Calif. 94720

University of California
Los Angeles, Calif. 90024

California State Polytechnic College
San Luis Obispo, Calif. 93401

Carnegie-Mellon University
Pittsburgh, Pa. 15213

METALS

Case Western Reserve University
Cleveland, Ohio 44106

University of Chicago
Chicago, Ill. 60637

University of Cincinnati
Cincinnati, Ohio 45221

Clemson University
Clemson, S.C. 29631

Cleveland State University
Cleveland, Ohio 44115

Colorado School of Mines
Golden, Colo. 80401

Columbia University
Henry Krumb School of Mines
New York, N.Y. 10027

University of Connecticut
Storrs, Conn. 06268

Cornell University
Ithaca, N.Y. 14850

University of Delaware
Newark, Del. 19711

University of Denver
Denver, Colo. 80210

Don Bosco Technical Institute
Rosemead, Calif. 91770

Drexel University
Philadelphia, Pa. 19104

University of Florida
Gainesville, Fla. 32601

Georgia Institute of Technology
Atlanta, Ga. 90332

Grove City College
Grove City, Pa. 16127

Harvard University
Cambridge, Mass. 02138

University of Idaho
Moscow, Idaho 83843

University of Illinois
Urbana, Ill. 61801

University of Illinois at Chicago
Circle
Chicago, Ill. 60680

Illinois Institute of Technology
Chicago, Ill. 60616

Iowa State University of Science
and Technology
Ames, Iowa 50010

University of Kentucky
Lexington, Ky. 40506

Lafayette College
Easton, Pa. 18042

Lehigh University
Bethlehem, Pa. 18015

Marquette University
Milwaukee, Wisc. 53233

University of Maryland
College Park, Md. 20742

University of Massachusetts
Amherst, Mass. 01002

Massachusetts Institute of
Technology
Cambridge, Mass. 02139

McGill University
Montreal, Que., Canada

McMaster University
Hamilton, Ont., Canada

University of Michigan
Ann Arbor, Mich. 48104

Michigan State University
East Lansing, Mich. 48823

Michigan Technological University
Houghton, Mich. 49931

Milwaukee Institute of Technology
Milwaukee, Wisc. 53203

University of Minnesota
Minneapolis, Minn. 55455

Mississippi State University
State College, Miss. 39762

University of Missouri–Rolla
Rolla, Mo. 65401

Montana College of Mineral Science
and Technology
Butte, Mont. 59701

University of Montreal Ecole
Polytechnique
Montreal, Que., Canada

University of Nebraska
Lincoln, Neb. 68508

University of Nevada
MacKay School of Mines
Reno, Nev. 89507

New Mexico Institute of Mining and
Technology
Socorro, N.M. 87801

New York University
New York, N.Y. 10003

State University of New York
Stony Brook, L.I., N.Y. 11790

North Carolina State University
Raleigh, N.C. 27607

Northwestern University
Evanston, Ill. 60201

University of Notre Dame
Notre Dame, Ind. 46556

Nova Scotia Technical College
Halifax, N.S., Canada

Ohio State University
Columbus, Ohio 43210

University of Oklahoma
Norman, Okla. 73069

Oregon State University
Corvallis, Ore. 97330

Pennsylvania State University
University Park, Pa. 16802

University of Pennsylvania
Philadelphia, Pa. 19104

University of Pittsburgh
Pittsburgh, Pa. 15213

Purdue University
Lafayette, Ind. 47907

Queen's University
Kingston, Ont., Canada

Rensselaer Polytechnic Institute
Troy, N.Y. 12181

Rice University
Houston, Tex. 77001

San Jose State College
San Jose, Calif. 95114

South Dakota School of Mines and
  Technology
Rapid City, S.D. 57701

Stanford University
Stanford, Calif. 94305

Stevens Institute of Technology
Hoboken, N.J. 07030

Syracuse University
Syracuse, N.Y. 13210

University of Tennessee
Knoxville, Tenn. 37916

University of Texas
Austin, Tex. 78712

University of Texas at El Paso
El Paso, Tex. 79999

University of Toronto
Toronto 5, Ont., Canada

U.S. Naval Post Graduate School
Monterey, Calif. 93940

University of Utah
Salt Lake City, Utah 84112

Vanderbilt University
Nashville, Tenn. 37203

University of Virginia
Charlottesville, Va. 22901

Virginia Polytechnic Institute
Blacksburg, Va. 24061

University of Washington
Seattle, Wash. 89105

Washington State University
Pullman, Wash. 99163

Wayne State University
Detroit, Mich. 48202

West Virginia University
Morgantown, W.Va. 26506

University of Wisconsin
Madison, Wisc. 53706

Yale University
New Haven, Conn. 06520

Youngstown State University
Youngstown, Ohio 44503

Federal Libraries: One might expect federal libraries to be generally available to the public, but this is not necessarily so. Many are operated as an adjunct to specialized information services or are limited to serving the needs of a particular scientific or research agency of the government. (See Chapter 4 on Federal Agencies and directory section C.)

Pricing of Library Services: Most library services are free (probably giving rise to the notion that *all* information resources should be provided at no charge), and this, of course, is a valid reason for suggesting them as a first resource. Special services, such as extensive literature searches, photocopy services, and postage on loans, are likely to be charged at cost.

TECHNICAL SOCIETIES

The technical society is one of the very best sources of discipline-oriented information. It brings this information to its members not only in the form of research results, data, and news about industry, but also serves as a pipeline of intercommunication. It is a good source to find out "who's doing what."

In this chapter, "technical society" is used as an omnibus term to include not only technical societies but also professional institutes (for which academic qualifications are required), research associations, and trade associations. In the listings at the end of this section, professional institutes and technical societies are combined, as are research associations and trade associations, because the general statements that can be made about the information resources of technical and professional societies differ from those that can be made about the information resources of research and trade associations. In the directory of associations, section B, all four are intermixed in alphabetical order.

Technical and professional societies make their information services and products widely available, but generally there is a charge for them. The member almost always has significant price privileges, and some information services—such as a society journal or news bulletin—may even be included, without extra charge, in the membership fee. A few publications, products, and services may be limited to members only, but these are the exception rather than the rule.

The situation with regard to trade associations is a variable one. Some are so dedicated to promoting the use of their products (sometimes a particular metal or alloy) that they will provide all sorts of free information to anyone requesting it. Others (a notable example being the American Iron and Steel Institute) restrict their services so rigidly to their member companies that they have asked not to be listed in this publication.

Pricing practices vary both among and between technical societies and trade associations. Most technical societies are open to individual memberships, and their dues are remarkably low when balanced against the services provided. It would be advantageous for an individual who has any prospect of using even a relatively small portion of the society's products or services to take out a membership (assuming he meets qualifications) rather than to buy the products independently."

Trade association membership is usually on an organization (industrial company) basis; therefore, the total membership is small and dues are high. However, the "general public" usually gets a bargain when it can make use of trade association materials. Bear in mind that many trade association publications will be promotional rather than technical and scientific; much of their information is of a commercial or statistical nature. Many are engaged in the writing of specifications, standards, or recommended practices, and this is another reason for not listing them with "technical" societies. On the other hand, some of them have highly developed technical information services, for example, The Aluminum Association (B2) and the Copper Development Association, Inc. (B31).

For additional information about associations in the United States, including number of members, chief executives, committees, meetings, and other activities, consult Gale's *Encyclopedia of Associations* (E7 and G8). For British associations, a good source is *Industrial Research in Britain* (G14).

TECHNICAL AND PROFESSIONAL SOCIETIES

American Ceramic Society (B4)
American Chemical Society (B5)
American Electroplaters' Society (B7)
American Foundrymen's Society (B8)
American Institute of Aeronautics and Astronautics (B10)
American Institute of Chemical Engineers (B11)
American Institute of Physics (B12)
American Powder Metallurgy Institute (B50)
American Society for Metals (B15)
American Society for Nondestructive Testing (B16)
American Society for Quality Control (B17)
American Society for Testing and Materials (B18)
American Society of Mechanical Engineers (B14)
American Vacuum Society (B19)
American Welding Society (B20)
Association of Iron and Steel Engineers (B21)
British Ceramic Society (B23)
Canadian Institute of Mining and Metallurgy (B27)
Canadian Welding Society (B28)
Electrochemical Society (B33)

Institute of British Foundrymen (B37)
Institute of Metal Finishing (B38)
Institute of Metals (B39)
Institution of Electrical Engineers (B40)
Institution of Mining and Metallurgy (B41)
Instrument Society of America (B42)
International Microstructural Analysis Society (B44)
Iron and Steel Institute (B46)
The Metallurgical Society, American Institute of Mining, Metallurgical and Petroleum Engineers (B52)
National Association of Corrosion Engineers (B53)
Society for Analytical Chemistry (B59)
Society for Experimental Stress Analysis (B63)
Society of Aerospace Material and Process Engineers (B58)
Society of Automotive Engineers (B60)
Society of Chemical Industry (B61)
Society of Die Casting Engineers (B62)
Society of Manufacturing Engineers (B64)
Society of Mining Engineers, American Institute of Mining, Metallurgical and Petroleum Engineers (B65)
The Welding Institute (B70)
Wire Association, Inc. (B71)

RESEARCH AND TRADE ASSOCIATIONS

Aluminium Federation (B1)
The Aluminum Association (B2)
Aluminum Smelters Research Institute (B3)
American Die Casting Institute (B6)
American Hot Dip Galvanizers Association (B9)
American Iron and Steel Institute
American Metal Stamping Association (B13)
British Cast Iron Research Association (B22)
British Iron and Steel Research Association (B24)
British Non-Ferrous Metals Research Association (B25)
Canadian Copper & Brass Development Association (B26)
Centre International de Développement de l'Aluminium (B29)
Copper Development Association (U.K.) (B30)
Copper Development Association, Inc. (New York) (B31)
Ductile Iron Society (B32)
Forging Industry Association (B34)
Gray and Ductile Iron Founders' Society (B35)
Industrial Heating Equipment Association (B36)

International Lead Zinc Research Organization (B43)
Investment Casting Institute (B45)
Lead Development Association (See ZDA/LDA Abstracting Service, F13)
Lead Industries Association (B47)
Malleable Founders' Society (B48)
Metal Powder Industries Federation (B50)
Metal Treating Institute (B51)
Non-Ferrous Founders' Society (B54)
Open Die Forging Institute (B55)
Production Engineering Research Association (B56)
Selenium-Tellurium Development Association, Inc. (B57)
Steel Castings Research and Trade Association (B66)
Steel Founders' Society of America (B67)
Tin Research Institute (B68)
Welded Steel Tube Institute (B69)
Zinc Development Association (See ZDA/LDA Abstracting Service, F13)
Zinc Institute (B72)

2

ABSTRACTS, INDEXES, BIBLIOGRAPHIES, REVIEWS

Abstracts, indexes, and bibliographies are secondary publications (secondary, according to one Webster definition, "having a derivative rank, position, or sequence") because they are derived from the primary publications. An *index* performs the first coarse screening function in leading a searcher to what he is trying to find. The *abstract* becomes a finer screen in discarding the dross. *Bibliographies* take the search a step further by grouping together references on specific topics. Finally, the searcher may find one or more important primary publications, or he may discover that the knowledge he seeks doesn't exist—which in itself is positive rather than negative information, particularly if he is contemplating a research project on the topic in question.

Reviews are borderline. Critical reviews can be considered as primary publications because they represent opinions and deductions of the author; but, because they also generally summarize, collate, and provide bibliographic data culled from numerous primary publications, they will be considered in this chapter as secondary in nature. Frequently, reviews are considered in a completely different category, namely, "tertiary publications."

ABSTRACTS AND INDEXES

Almost all abstracting publications include indexes, and some indexes include abstracts (*Engineering Index* is a good example), therefore, these will be combined. A list of the principal English-

11

language abstracting and indexing publications that deal with metals to some degree is given on pages 13–16, together with directory numbers for source and further detail. The FID directory of *Abstracting Services* (G1) was a primary source for much of the information concerning the services listed. (A number of technical societies and similar organizations publish abstracts of, or indexes to, their own publications only. These are not shown in the list in this chapter, but are noted in the directory descriptions.)

There is only one abstract journal in the English language that covers all aspects of metallurgy, exclusively, in a single publication, and that is *Metals Abstracts* (and its companion publication *Metals Abstracts Index*). The only comparable publications in other languages are: *Bulletin Signalétique*, 740, *Métaux—Métallurgie* in French, published by the Centre Nationale de la Recherche Scientifique in Paris; and *Referativnyi Zhurnal-Metallurgiya* in Russian, published by VINITI in Moscow.

A number of abstract journals are devoted to specific metals—aluminum, copper, zinc, lead, etc.—or to metallurgical processes or phenomena—corrosion, steel casting, metal forming. Their coverage is clearly indicated by their titles. Other journals include some aspects of metallurgy as part of a broader coverage (*Chemical Abstracts, Physics Abstracts, Nuclear Science Abstracts*, for example); the extent of metallurgical coverage is defined in the directory descriptions. Nevertheless, it is almost impossible to indicate in one or two sentences which aspects of chemical metallurgy that might be found in *Chemical Abstracts* will not be covered in *Metals Abstracts*. Only a skilled metals librarian will know which source to try first.

Metals Abstracts was formed in 1968 by the merger of two existing abstract journals, *Review of Metal Literature* (RML) and *Metallurgical Abstracts* (*Met.Abs.*); *Met.Abs.* was established in 1908 and *RML* in 1944. They differed significantly in content and style; *Met.Abs.* covered primarily nonferrous literature (ferrous literature was covered by the British Iron and Steel Institute) and concentrated primarily on scientific metallurgy as opposed to the practical metallurgy of everyday plant operations; *RML* covered all metals, with emphasis on operating metallurgy (scientific and physical metallurgy were included, but the pure physics of metals was not covered as thoroughly as in *Met.Abs.*). *Met.Abs.* published informative—sometimes even critical—abstracts, whereas *RML* published brief indicative abstracts (annotations). In 1965, *RML* began to publish informative abstracts, and *Met.Abs.* expanded its coverage to include ferrous metallurgy. It was obvious they were on a collision course; thus, the merger to form *Metals Abstracts.*

The directory descriptions differentiate the index-type publications, which carry bibliographic citations only, from abstract-type publications, which carry amplifying data concerning the content of the documents. Most index-type publications have supplemental author indexes—either in each issue or in an annual issue. Most abstract-type publications issue annual subject and author indexes, in addition to displaying the abstracts under broad subject headings; a few carry subject and/or author indexes in each issue. These data are given in the individual descriptions.

Index-type publications tend to be more current than abstract journals, particularly when they are computer-produced; they serve more as an "alerting" tool than as an archival reference source. Index-type publications are listed separately below; all others will give "abstract" information of some sort. When an index publication is a companion to an abstract publication and is not intended to be used alone, it is listed with the abstract journal, for example, *Metals Abstracts* and *Metals Abstracts Index.*

ABSTRACT JOURNALS
(Some With Companion Index Publications)

Abstracts of Current Literature, Iron and Steel Institute (B46)

Aluminium Abstracts, Centre International de Développement de l'Aluminium (B29)

Analytical Abstracts, Society for Analytical Chemistry (B59)

Applied Mechanics Reviews, American Society of Mechanical Engineers (B14)

BCIRA Abstracts of Foundry Literature, British Cast Iron Research Association (B22)

Bibliographical Bulletin for Welding and Allied Processes, Institute de Soudure (F7)

British Ceramics Abstracts, British Ceramic Society (B23)

British Technology Index, The Library Association (F8)

Central Patents Index Alerting Bulletin M—Metallurgy, Derwent Publications Ltd. (E4)

Ceramic-Metal Systems and Enamel Bibliography and Abstracts, American Ceramic Society (B4)

Ceramic Abstracts, American Ceramic Society (B4)

Chemical Abstracts, Chemical Abstracts Service (F3)

Cobalt—Review of Technical Literature, Cobalt Information Center (D7)

Copper Abstracts, Copper Development Association (U.K.) (B30)

Corrosion Abstracts, National Association of Corrosion Engineers (B53)

Corrosion Control Abstracts, Scientific Information Consultants Ltd. (E17)

Current Abstracts of Chemistry and Index Chemicus, Institute for Scientific Information (E10)

Current Literature on Nondestructive Testing, American Society for Nondestructive Testing (B16)

Current Physics Advance Abstracts, American Institute of Physics (B12)

Czechoslovak Science and Technology Digest, Scientific Information Consultants Ltd. (E17)

Dissertation Abstracts, University Microfilms (E20)

East European Scientific Abstracts, Materials Science and Metallurgy, Joint Publications Research Service (C3)

Engineering Index, Engineering Index, Inc. (F4)

Extracts of Documents on Copper Technology, and *Coordinate Index to Information and Data on Copper Technology*, Copper Development Association, Inc. (B31 and D8)

Gold Bulletin, Chamber of Mines of South Africa (F2)

Government Reports Announcements, and *Government Reports Index*, National Technical Information Service (C10)

Hungarian Technical Abstracts, Hungarian Central Technical Library and Documentation Centre (F6)

IMM Abstracts, Institution of Mining and Metallurgy (B41)

International Aerospace Abstracts, American Institute of Aeronautics and Astronautics (B10)

Journal of Applied Chemistry—Abstracts, Society of Chemical Industry (B61)

Lead Abstracts, ZDA/LDA Abstracting Service (F13)

Liquid Metals Technology Abstract Bulletin, MSA Research Corp. (E13)

Mechanical Sciences Abstracts, Scientific Information Consultants Ltd. (E17)

Metal Finishing Abstracts, Finishing Publications Ltd. (E6)

Metalforming Digest, American Society for Metals (B15)

Metallurgical Abstracts, Institute of Metals (B39)

Metallurgical Abstracts on Light Metals and Alloys, Light Metal Educational Foundation, Inc. (F9)

Metal Powder Report, Powder Metallurgy Ltd. (E16)

Metals Abstracts, and *Metals Abstracts Index,* American Society for Metals (B15) and Institute of Metals (B39)

Nuclear Science Abstracts, U.S. Atomic Energy Commission (C15)

Official Gazette of the United States Patent Office, U.S. Patent Office (C17)

PERA Bulletin, Production Engineering Research Association (B56)

Physics Abstracts, Institution of Electrical Engineers (B40)

Plating—Patents Abstracts, American Electroplaters' Society (B7)

Platinum Metals Review, Johnson Matthey & Co. Ltd. (E11)

Powder Metallurgy Science & Technology, Franklin Institute (F5)

Review of Metal Literature, American Society for Metals (B15)

Science Abstracts, Institution of Electrical Engineers (B40)

Scientific and Technical Aerospace Reports, National Aeronautics and Space Administration (C5)

Selenium and Tellurium Abstracts, Selenium-Tellurium Development Association, Inc. (B57)

Steel Casting Abstracts, Steel Castings Research and Trade Association (B66)

Technical Abstract Bulletin, and *Technical Abstract Bulletin Index,* Defense Documentation Center (C2)

U.S. Government Research and Development Reports, and *U.S. Government Research and Development Reports Index,* National Technical Information Service (C10)

USSR Scientific Abstracts, Materials Science and Metallurgy, Joint Publications Research Service (C3)

Vacuum—Classified Abstracts, Pergamon Press, Inc. (E14)

Weekly Government Abstracts, National Technical Information Service (C10)

World Aluminum Abstracts, The Aluminum Association (B2)

Zinc Abstracts, ZDA/LDA Abstracting Service (F13)

INDEX PUBLICATIONS

Aeronautical Engineering—A Special Bibliography With Indexes, National Aeronautics and Space Administration (C5)

American Doctoral Dissertations, University Microfilms (E20)

Applied Science & Technology Index, H. W. Wilson Co. (E21)

Bibliographic Index, H. W. Wilson Co. (E21)

Chemical Abstracts Service Source Index, Chemical Abstracts Service (F3)

Chemical Titles, Chemical Abstracts Service (F3)

Current Contents, Institute for Scientific Information (E10)

Current Papers in Physics (also *Current Papers in Electrical and Electronics Engineering* and *Current Papers on Computers and Control*), Institution of Electrical Engineers (B40)

Current Physics Titles, American Institute of Physics (B12)

Design, Engineering Materials & Hydraulic Drives Abstracts, Scientific Information Consultants Ltd. (E17)

Dissertation Digest, University Microfilms (E20)

Electronic Properties of Materials—A Guide to the Literature, Electronic Properties Information Center (D14)

Index to Forthcoming Russian Books, Scientific Information Consultants Ltd. (E17)

Index to Publications Related to Aluminum Extrusions, The Aluminum Association (B2)

Masters Theses in the Pure and Applied Sciences Accepted by Colleges and Universities of the United States, Thermophysical Properties Research Center (D30)

Monthly Catalog, U.S. Government Publications, Superintendent of Documents, U.S. Government Printing Office (C12)

Permuted Materials Index and *Author Index*, Alloy Data Center (D4)

Permuterm Subject Index, Institute for Scientific Information (E10)

Science Citation Index, Institute for Scientific Information (E10)

Sci/Tech Quarterly Index to U.S. Government Translations, CCM Information Corporation (E3)

Thermophysical Properties Research Literature Retrieval Guide, Thermophysical Properties Research Center (D30)

Titles of Current Literature, Iron and Steel Institute (B46)

Transdex—Guide to U.S. Government JPRS Translations of Iron-Curtain Documents, CCM Information Corporation (E3)

Translations Register-Index, John Crerar Library (A7)

Pricing: The so-called "information explosion" has followed an exponential curve over the past 25 years. The cost of secondary services has increased even more dramatically in this time. For example, the subscription price of *Chemical Abstracts* increased from $12 to $1,950 (although the section on *Applied Chemistry and Chemical Engineering* is a mere $70). *Review of Metal Literature* cost only $5 as late as 1964. Major changes—informative instead of indicative abstracts, greater coverage, monthly indexes—all forced the price upward to the present $455 for *Metals Abstracts* and its companion *Metals Abstracts Index*. Members of technical societies usually get a good price break, and ASM members can have the two publications for $42, but only if their organization also subscribes at the basic rate. Public and university libraries can get the combination subscriptions for $275. *Engineering Index*, at $450, is in about the same range.

Some trade associations provide their abstracts free "to qualified requesters" (a rather ambiguous restriction). Examples are *Lead Abstracts* and *Zinc Abstracts*. The Aluminum Association provides *World Aluminum Abstracts* free to its members, but the price to nonmembers is $75.

Commercial publishers usually set a fairly high price. *Science Citation Index* sells for $1,500, but the citation method of indexing does give it great extent of coverage over the years.

Government, the initiator of many free services in the past, now generally sets a charge, although it is usually minimal. *Nuclear Science Abstracts* costs $42, and its cumulative indexes, $38 per year. NASA's *Scientific and Technical Aerospace Reports* (STAR) costs $54, and its indexes, $30 per year.

BIBLIOGRAPHIES AND REVIEWS

To list all of the bibliographies that are available on metallurgical topics would require more space than this publication can afford. Even in the directory descriptions of the publications, products, and services of individual organizations, it was not practical to include all of the available titles; for example, currently there are 33 titles in the ASM Bibliography Series, with more being added constantly.

Following is a list of organizations that produce bibliographies. This list shows only those organizations that produce bibliographies in quantity; it does not include information centers and retrieval services that produce bibliographies on demand to user specifications (covered in Chapter 5). The directory descriptions will give some idea of type and content; a list of specific titles is usually available on request.

American Institute of Physics (B12)
American Society for Metals (B15)
American Society for Testing and Materials (B18)
British Cast Iron Research Association (B22)
Defense Documentation Center (C2)
Engineering Societies Library (A6)
Franklin Institute (F5)
Iron and Steel Institute (B46)
Linda Hall Library (A9)
Metal Powder Industries Federation (B50)
Nondestructive Testing Information Analysis Center (D22)
Rare-Earth Information Center (D24)
U.S. Atomic Energy Commission (C15)

Reviews are something of a rarity because of the large expense of compiling them and the difficulty in finding qualified authors who are willing to spend the necessary time to scan and analyze the literature.

One of the best series is *International Metallurgical Reviews*, published jointly by the Institute of Metals and the American Society for Metals (formerly *Metallurgical Reviews*, published by the Institute of Metals, 1966–1971). Many good review articles are buried in the journal literature, and someone could do a real service for the metallurgical community by compiling a "Bibliography of Reviews" such as the DDC *Bibliography of Bibliographies* (C2). The following probably is not a very complete list, but it does show a few of the more prominent review series.

Chemical Reviews, American Chemical Society (B5)
Diffusion Data, Diffusion Information Center (E5)
ILZRO Research Digest, International Lead Zinc Research Organization (B43)
International Metallurgical Reviews, Institute of Metals (B39), published jointly with the American Society for Metals (B15)
Progress in and *Annual Review* series, Pergamon Press, Inc. (E14)
Review of Iron and Steel Literature, Carnegie Library of Pittsburgh (A5)
Reviews of Recent Developments, Defense Metals and Ceramics Information Center (D11)

3

JOURNALS, BOOKS, TRANSLATIONS, STANDARDS, PATENTS

Journals and books are the fundamental vehicles for transfer of information via the printed word. They are indeed the true "primary publications." However, translations, standards, and patents, even though they are not always provided in the usual printed or published form, are also considered "primary publications," because they too convey original information.

Government reports also are primary resources of metallurgical information. These are covered in Chapter 4 on Federal Agencies.

Trade literature and company catalogs are often highly useful, but their value may be short-lived, and their technical information content highly variable. Therefore, no effort is made to list titles or names of companies that freely distribute their technical literature. Many libraries build quite extensive collections of trade literature and subject-catalog them in sufficient detail to make effective searches. Alternatively, the metallurgist can build his own collection.

JOURNALS

There are more than 1,000 worldwide journals and serials that publish at least occasional articles or papers of metallurgical interest. A reprint of the list as it appears in the most recent annual volume of *Metals Abstracts*, showing country of origin but not complete address, is available from the American Society for Metals on request (B15).

Other abstracting services publish similar lists, for example, *Publications Indexed for Engineering* (available from Engineering Index, Inc., F4). The *Chemical Abstracts Service Source Index* (F3) is by far the most comprehensive and informative list covering scientific and technical serials, but its price of $100, plus a $75 charge for quarterly updating, tends to limit it to library shelves. The same is true of *Ulrich's International Periodicals Directory*,[1] probably the most widely used resource for journals in all fields.

Most of the 1,000-plus titles in the *Metals Abstracts* list give only partial coverage to metallurgy; in my opinion, only 103 among those in the English language are of prime, or almost exclusive, metallurgical interest. The basic criterion for selection was "traffic" through the journals for *Metals Abstracts*, which means the number of items selected for abstracting as compared to the total number published—sometimes known in the trade as abstract "productivity" of a journal. The list contains a selection of both scientific and technological journals. It is confined to English-language publications, but does include a number of Russian journals that are translated cover-to-cover and a few journals published in Japan but written in English.

The *Guide to Metallurgical Information* (G11), published by the Special Libraries Association in 1965, contains a similar list of serials with 96 titles in the English language (which differs considerably from the author's list of 103). This book also gives the important serials of 10 foreign countries. Following is a list of 103 prime metallurgical journals.

Acta Metallurgica
Pergamon Press, Inc.
Maxwell House, Fairview Park
Elmsford, N.Y. 10523

Alloy Digest
Engineering Alloys Digest, Inc.
P.O. Box 156
Upper Montclair, N.J. 07043

Anti-Corrosion Methods and Materials
Sawell Publications Ltd.
4 Ludgate Circus
London, E.C.4, England

Automatic Welding (USSR)
(English translation of
Avtomaticheskaya Svarka)
The Welding Institute
Abington Hall
Abington, Cambridge, England

Blast Furnace and Steel Plant
Steel Publications, Inc.
624 Grant Bldg.
Pittsburgh, Pa. 15230

British Corrosion Journal
British Joint Corrosion Group
14 Belgrave Sq.
London, S.W.1, England

[1] *Ulrich's International Periodicals Directory.* R. R. Bowker Co., 1180 Avenue of the Americas, New York, N.Y. 10036. Two volumes. 1932– . Biennial. $34.50/set.

British Foundryman
Institute of British Foundrymen
137-139 Euston Rd.
London, N.W.1, England

*British Journal of Non-Destructive
    Testing*
Non-Destructive Testing Society of
    Great Britain
Maitland House
Warrior Sq.
Southend-On-Sea
Essex, SS1 2JY, England

*Bulletin of the Japan Society of
    Mechanical Engineers*
Nihon Kikaku Kyokai Bldg.
1-24, 4-chome, Akasaku
Minato-ku
Tokyo, Japan

Canadian Metallurgical Quarterly
Canadian Institute of Mining and
    Metallurgy
1117 St. Catherine St , W.
Montreal 110, Que., Canada

Cast Metals Research Journal
American Foundrymen's Society
Golf and Wolf Rds.
Des Plaines, Ill. 60016

Cobalt (English Edition)
Cobalt Information Center
Battelle Memorial Institute
505 King Ave.
Columbus, Ohio 43201

Composites
IPC Science & Technology Press Ltd.
IPC House
32 High St.
Guildford, Surrey, England

Corrosion
National Association of Corrosion
    Engineers
2400 W. Loop S.
Houston, Tex. 77027

Corrosion Prevention and Control
Scientific Surveys Ltd.
11A Gloucester Rd.
London, S.W.7, England

Corrosion Science
Pergamon Press, Inc.
Maxwell House, Fairview Park
Elmsford, N.Y. 10523

Crystal Lattice Defects
Gordon and Breach Science
    Publishers Ltd.

12 Bloomsbury Way
London, W.C.1, England

Die Casting Engineer
Society of Die Casting Engineers,
    Inc.
16007 W. Eight Mile Rd.
Detroit, Mich. 48235

Electroplating and Metal Finishing
167 Hagden Lane
Watford, WD1 8LW, England

Engineering Fracture Mechanics
Pergamon Press, Inc.
Maxwell House, Fairview Park
Elmsford, N.Y. 10523

Experimental Mechanics
Society for Experimental Stress
    Analysis
21 Bridge Sq.
Westport, Conn. 06880

Foundry
Penton Publishing Co.
Penton Plaza
Cleveland, Ohio 44114

Giessereiforschung in English
A. S. Morrison
Penlee, Thirlstane Rd.
Malvern, Worcestershire, England

Industrial Finishing (U.S.)
Hitchcock Publishing Co.
Hitchcock Bldg.
Wheaton, Ill. 60187

Industrial Heating
National Industrial Publishing Co.
Union Trust Bldg.
Pittsburgh, Pa. 15219

Industrial Laboratory (USSR)
(English translation of *Zavodskaya
    Laboratoriya*)
Consultants Bureau
227 W. 17th St.
New York, N.Y. 10011

*International Journal of Fracture
    Mechanics*
Wolters-Noordhoff Publishing
Box 58
Groningen, The Netherlands

*International Journal of Powder
    Metallurgy*
American Powder Metallurgy
    Institute
201 E. 42nd St.
New York, N.Y. 10017

Iron and Steel
IPC Science and Technology Press Ltd.
IPC House
32 High St.
Guildford, Surrey, England

Iron and Steel Engineer
Assn. of Iron and Steel Engineers
Empire Bldg.
Pittsburgh, Pa. 15222

Journal of Composite Materials
Technomic Publishing Co., Inc.
750 Summer St.
Stamford, Conn. 06901

Journal of Electronmicroscopy
Japanese Society of Electron Microscopy
334, 3-chome Hyakunin-cho
Shinjuku-ku, Tokyo, Japan

Journal of Materials
American Society for Testing and Materials
1916 Race St.
Philadelphia, Pa. 19103

Journal of Materials Science
Chapman and Hall
11 New Fetter Lane
London, E.C.4, England

Journal of Metals
American Institute of Mining, Metallurgical and Petroleum Engineers
345 E. 47th St.
New York, N.Y. 10017

Journal of Physics and Chemistry of Solids
Pergamon Press, Inc.
Maxwell House, Fairview Park
Elmsford, N.Y. 10523

Journal of the Australian Institute of Metals
23 McKillop St.
Melbourne, C.1, Australia

Journal of the Electrochemical Society
30 E. 42nd St.
New York, N.Y. 10017

Journal of the Institute of Metals
17 Belgrave Sq.
London, S.W.1, England

Journal of the Iron and Steel Institute
1, Carlton House Terr.
London, S.W.1, England

Journal of the Less-Common Metals
Elsevier Sequoia SA
Box 851
1001 Lausanne 1, Switzerland

Light Metal Age
693 Mission St.
San Francisco, Calif. 94105

Materials Engineering
600 Summer St.
Stamford, Conn. 06904

Materials Evaluation
American Society for Nondestructive Testing
914 Chicago Ave.
Evanston, Ill. 60202

Materials Protection and Performance
National Association of Corrosion Engineers, Inc.
2400 W. Loop S.
Houston, Tex. 77027

Materials Research and Standards
American Society for Testing and Materials
1916 Race St.
Philadelphia, Pa. 19103

Materials Science and Engineering
Elsevier Sequoia SA
Box 851
1001 Lausanne 1, Switzerland

Metal Construction & British Welding Journal
The Welding Institute
Abington Hall
Abington, Cambridge, England

Metal Finishing
Metals & Plastics Publications, Inc.
99 Kinderkamack Rd.
Westwood, N.J. 07675

Metal Finishing Journal
Fuel and Metallurgical Journals Ltd.
John Adam House
17-19 John Adam St., Adelphi
London, W.C.2, England

Metallurgia and Metal Forming
Fuel and Metallurgical Journals Ltd.
John Adam House
17-19 John Adam St., Adelphi
London, W.C.2, England

Metallurgical Transactions
Metallurgical Society of AIME, and American Society for Metals
345 E. 47th St.
New York, N.Y. 10017

Metallurgist (USSR)
(English translation of *Metallurg*)
Consultants Bureau
227 W. 17th St.
New York, N.Y. 10011

Metal Progress
American Society for Metals
Metals Park, Ohio 44073

Metals and Materials
Institute of Metals
17 Belgrave Sq.
London, S.W.1, England

Metal Science and Heat Treatment
(USSR)
(English translation of *Metallove-denie i Termicheskaya Obrabotka Metallov*)
Consultants Bureau
227 W. 17th St.
New York, N.Y. 10011

Metal Science Journal
Institute of Metals
17 Belgrave Sq.
London, S.W.1, England

Metals Engineering Quarterly
American Society for Metals
Metals Park, Ohio 44073

Metal Treating
Temple Publications, Inc.
Box 1386
Rocky Mount, N.C. 27801

Mitsubishi Heavy Industries Technical Review
Mitsui Shipbuilding & Engineering
Co. Ltd.
6-4 Tsukiji 5-chome
Chuo-ku, Tokoyo, Japan

Modern Casting
American Foundrymen's Society
Golf & Wolf Rds.
Des Plaines, Ill. 60016

Modern Metals
919 N. Michigan Ave.
Chicago, Ill. 60611

Non-Destructive Testing
IPC Science & Technology Press Ltd.
32 High St.
Guildford, Surrey, England

Oxidation of Metals
Plenum Publishing Corp.
227 W. 17th St.
New York, N.Y. 10011

Philosophical Magazine
Taylor & Francis Ltd.
10-14 Macklin St.
London, WC2B 5NX, England

Physics of Metals and Metallography
(USSR)
(English translation of *Fizika Metallov i Metallovedenie*)
Pergamon Press, Inc.
Maxwell House, Fairview Park
Elmsford, N.Y. 10523

Plating
American Electroplaters' Society, Inc.
56 Melmore Gardens
E. Orange, N.J. 07017

Platinum Metals Review
Johnson, Matthey & Co. Ltd.
78 Hatton Garden
London, EC1P IAE, England

Powder Metallurgy
Powder Metallurgy Joint Group
Iron & Steel Institute and Institute
of Metals
17 Belgrave Sq.
London, S.W.1, England

Powder Metallurgy International
Verlag Schmid GmbH
Kaiser-Joseph-Str. 217
P.O. Box 1722
Freiburg/Brsg., Germany

Powder Technology
Elsevier Sequoia SA
Box 851
1001 Lausanne 1, Switzerland

Praktische Metallographie—Practical Metallography (in English and German)
Postfach 447
D-7000, Stuttgart 1, Germany

Precision Metal
Industrial Publishing Co.
614 Superior Ave.
Cleveland, Ohio 44113

Products Finishing (Cincinnati)
Gardner Publications, Inc.
600 Main St.
Cincinnati, Ohio 45202

Progress in Materials Science
Pergamon Press Ltd.
Headington Hill Hall
Oxford, England

Protection of Metals (USSR)
(English translation of *Zashchita Metallov*)
Plenum Publishing Corp.
227 W. 17th St.
New York, N.Y. 10011

Russian Castings Production
(English translation of *Liteinoe Proizvodstvo*)
British Cast Iron Research Association
Bordesley Hall
Alvechurch, Birmingham, England

Russian Metallurgy (*Metally*)
(English translation of *Izvestiya Akademii Nauk SSSR, Metally*)
Scientific Information Consultants Ltd.
661 Finchley Rd.
London, NW2 2HN, England

SAMPE Journal
Society of Aerospace Material and Process Engineers
Technology Publishing Corp.
825 S. Barrington Ave.
Los Angeles, Calif. 90049

SAMPE Quarterly
Society of Aerospace Material and Process Engineers
P.O. Box 613
Azusa, Calif. 91702

Scripta Metallurgica
Pergamon Press, Inc.
Maxwell House, Fairview Park
Elmsford, N.Y. 10523

Soviet Journal of Non-Ferrous Metals
(English translation of *Tsvetnye Metally*)
Primary Sources
11 Bleeker St.
New York, N.Y. 10012

Soviet Materials Science
(English translation of *Fiziko-Khimicheskaya Mekhanika Materiallov*)
Faraday Press
84 Fifth Ave.
New York, N.Y. 10011

Soviet Physics—Crystallography
(English translation of *Kristallografiya*)
American Institute of Physics

335 E. 45th St.
New York, N.Y. 10017

Soviet Physics—Solid State
(English translation of *Fizika Tverdogo Tela*)
American Institute of Physics
335 E. 45th St.
New York, N.Y. 10017

Soviet Powder Metallurgy and Metal Ceramics
(English translation of *Poroshkovaya Metallurgiya*)
Consultants Bureau
227 W. 17th St.
New York, N.Y. 10011

Steel in the USSR
(Selected translations from the Russian of *Stal'* and *Izvestiya VUZ Chernaya Metallurgiya*)
Iron and Steel Institute
1, Carlton House Terr.
London, S.W.1, England

Sumitomo Search
31, Kawaramachi 4-chome
Higashi-ku
Osaka, Japan

Surface Science
North-Holland Publishing Co.
P.O. Box 3489
Amsterdam, The Netherlands

Thin Films
Gordon & Breach Science Publishers Ltd.
8 Bloomsbury Way
London, W.C.1, England
and
150 Fifth Ave.
New York, N.Y. 10011

Thin Solid Films
Elsevier Sequoia SA
Box 851
1001 Lausanne 1, Switzerland

Thirty-Three Magazine
22 Bank St.
Summit, N.J. 07901

Tin and Its Uses
Tin Research Institute
Fraser Rd.
Greenford, Middlesex, England

Transactions of National Research Institute for Metals (*Japan*)
300, 2-chome
Nakameguro, Meguro-ku
Tokyo, Japan

*Transactions of the Institute of
Metal Finishing*
178 Goswell Rd.
London, E.C.1, England

*Transactions of the Iron and Steel
Institute of Japan*
Keidanren Kaikan
9-4 Otemachi 1-chome
Chiyoda-ku
Tokyo, Japan

*Transactions of the Japan Institute
of Metals*
165, 3-chome
Omachi
Sendai, Japan

Wear
Elsevier Sequoia SA
Box 851
1001 Lausanne 1, Switzerland

Welding and Metal Fabrication
Engineering Chemical & Marine
Press Ltd.

Dorset House
Stamford St.
London, S.E.1, England

Welding Engineer
Welding Engineer Publications, Inc.
Box 128
Morton Grove, Ill. 60053

Welding Journal
American Welding Society
2501 N.W. 7th St.
Miami, Fla. 33125

Welding Production (USSR)
(English translation of *Svarochnoe
Proizvodsto*)
The Welding Institute
Abington Hall,
Abington, Cambridge, England

Wire Journal
209 Montowese St.
Branford, Conn. 06405

BOOKS

In 1962, the ASM Advisory Committee on Metallurgical Education compiled a "Metallurgical Book List" containing 40 of the most popular titles. In order to bring this list up to date, it was submitted to three reviewers (heads of metallurgical departments of major engineering universities). The following list of basic books on metallurgy consolidates their recommendations.[2]

Alloy Phase Equilibria. A. Prince. New York, N.Y., American
Elsevier, 1966.
An Introduction to Metallurgy. A. H. Cottrell. New York, N.Y.,
St. Martin's, 1967.
Constitution of Binary Alloys. 2nd ed., M. Hansen, 1958; 1st
Suppl., R. P. Elliott, 1965; 2nd Suppl., F. A. Shunk, 1969. New
York, N.Y., McGraw-Hill.
Diffusion in Solids. Paul G. Shewmon. New York, N.Y., McGraw-
Hill, 1963.
Elements of Materials Science. Lawrence H. Van Vlack. Reading,
Mass., Addison-Wesley, 1964.

[2] The list covers basic physical and mechanical metallurgy, structure and metallography, and theoretical metallurgy. A few recommended titles covering metallurgical specialties such as corrosion, foundry practice, and extractive metallurgy are omitted.

Elements of Physical Metallurgy. A. G. Guy. Reading, Mass., Addison-Wesley, 1959.

Engineers Guide to High-Temperature Materials. F. J. Clauss. Reading, Mass., Addison-Wesley, 1969.

Fracture of Structural Materials. A. S. Tetelman and A. J. McEvily. New York, N.Y., Wiley, 1967.

Interpretation of Metallographic Structures. W. Rostoker and J. R. Dvorak. New York, N.Y., Academic Press, 1965.

Introduction to Physical Metallurgy. S. H. Avner. New York, N.Y., McGraw-Hill, 1964.

Making, Shaping and Treating of Steel. Pittsburgh, Pa., U.S. Steel Corp., 1971.

Man, Metals and Modern Magic. J. G. Parr. Metals Park, Ohio, American Society for Metals, 1958.

Mechanical Behavior of Materials. F. A. McClintock and A. S. Argon. Reading, Mass., Addison-Wesley, 1966.

Mechanical Metallurgy. G. E. Dieter. New York, N.Y., McGraw-Hill, 1961.

Mechanical Properties of Matter. A. H. Cottrell. New York, N.Y., Wiley, 1964.

Metallic Materials in Engineering. Carl H. Samans. New York, N.Y., Macmillan, 1963.

Metallurgical Engineering. R. Schuhmann. Reading, Mass., Addison-Wesley, 1951.

Modern Metallography. R. E. Smallman and K. H. G. Ashbee. New York, N.Y., Pergamon, 1966.

Phase Diagrams in Metallurgy. F. N. Rhines. New York, N.Y., McGraw-Hill, 1956.

Phase Stability in Metals and Alloys. P. S. Rudman, J. Stringer, and R. I. Jaffee. New York, N.Y., McGraw-Hill, 1967.

Phase Transformations (Seminar). Metals Park, Ohio, American Society for Metals, 1970.

Physical Chemistry of Metals. L. S. Darken and R. W. Gurry. New York, N.Y., McGraw-Hill, 1953.

Physical Examination of Metals. Bruce Chalmers and A. G. Quarrell. New York, N.Y., St. Martin's, 1960.

Physical Metallurgy. R. W. Cahn. New York, N.Y., Wiley, 1965.

Physical Metallurgy for Engineers. A. G. Guy. Reading, Mass., Addison-Wesley, 1962.

Physical Metallurgy Principles. Robert E. Reed-Hill. New York, N.Y., Van Nostrand, 1964.

Physics of Solids. Charles A. Wert and Robb M. Thompson. New York, N.Y., McGraw-Hill, 1964.

Plastic Deformation of Metals. R. W. Honeycombe. New York, N.Y., St. Martin's, 1968.

Principles of Metallographic Laboratory Practice. G. L. Kehl. New York, N.Y., McGraw-Hill, 1949.

Recrystallization, Grain Growth and Textures (Seminar). Metals Park, Ohio, American Society for Metals, 1965.

Selected Values of Thermodynamic Properties of Metals and Alloys. R. Hultgren, R. L. Orr, P. D. Anderson, and K. K. Kelley. New York, N.Y., Wiley, 1963.

Strengthening Mechanisms in Solids (Seminar). Metals Park, Ohio, American Society for Metals, 1962.

Structure and Properties of Alloys. R. M. Brick, Robert B. Gordon, and Arthur Phillips. New York, N.Y., McGraw-Hill, 1965.

Structure of Metals. C. S. Barrett and T. Massalaskie. New York, N.Y., McGraw-Hill, 1966.

Structure of Metals and Alloys. William Hume-Rothery, R. E. Smallman, and C. W. Haworth. London, Institute of Metals, 1969.

The Engineers Guide to Steel. A. Hanson and J. G. Parr. Reading, Mass., Addison-Wesley, 1965.

Theoretical Structural Metallurgy. A. H. Cottrell. New York, N.Y., St. Martin's, 1955.

Thermodynamics of Alloys. J. Wagner. Reading, Mass., Addison-Wesley, 1952.

Thermodynamics of Solids. Richard A. Swalin. New York, N.Y., Wiley, 1963.

Transformations in Metals. Paul G. Shewmon. New York, N.Y., McGraw-Hill, 1969.

The reader is referred once again to the SLA *Guide* (G11) for extensive coverage of metallurgical books and reference volumes. It has the added advantage of arrangement by subject and inclusion of foreign publications.

Handbooks and similar compendia are of sufficient importance to warrant an updated list here. Title, editor (if any), and publisher are given in the list that follows. For addresses, see the numbered directory entry or the list of commercial book publishers. All titles on the list are current within the 1960–70 decade. For earlier titles still extant, or for more detail on dates, pages, and content, consult either the SLA *Guide* or your librarian.

Aerospace Structural Metals Handbook, Mechanical Properties Data Center (D19).

AIP Handbook, American Institute of Physics (B12), New York, N.Y., McGraw-Hill.

ASME Handbooks, American Society of Mechanical Engineers, (B14), New York, N.Y., McGraw-Hill.
>   *Metals Properties*
>   *Metals Engineering: Design*
>   *Metals Engineering: Processes*
>   *Engineering Tables*

British and Foreign Specifications for Steel Castings, Steel Castings Research and Trade Association (B66).

Casting Design Handbook, American Society for Metals (B15) and the U.S. Air Force.

Cast Metals Handbook, American Foundrymen's Society (B8).

Electroplating Engineering Handbook, A. K. Graham, ed., New York, N.Y., Reinhold.

Encyclopedia of Engineering Materials and Processes, H. R. Clauser, ed., New York, N.Y., Reinhold.

Forging Design Handbook, American Society for Metals (B15).

Foundry Sand Handbook, American Foundrymen's Society (B8).

Handbook of Chemistry and Physics, Cleveland, Ohio, Chemical Rubber Publishing Co.

Handbook of Electronic Materials, Electronic Properties Information Center (D14), New York, N.Y., Plenum Press.

Handbook of Experimental Stress Analysis, Society for Experimental Stress Analysis (B63).

Handbook of Industrial Electroplating, E. A. Ollard and E. B. Smith, New York, N.Y., American Elsevier Publishing Co.

Handbook of the Metallurgy of Tin, D. V. Belyayev. Elmsford, N.Y., Pergamon.

Handbook of Welded Steel Tubing, Welded Steel Tube Institute (B69).

Handbook of Welding Design, The Welding Institute (B70).

Investment Casting Handbook, Investment Casting Institute (B45).

Machining Data Handbook, Machinability Data Center (D3).

Malleable Iron Castings Handbook, Malleable Founders' Society (B48).

Materials Handbook, G. S. Brady, New York, N.Y., McGraw-Hill.
Metals Handbook, American Society for Metals (B15).
    Vol. 1—Properties and Selection of Metals
    Vol. 2—Heat Treating, Cleaning and Finishing
    Vol. 3—Machining
    Vol. 4—Forming
    Vol. 5—Forging and Casting
    Vol. 6—Welding
    Vol. 7—Atlas of Microstructures of Industrial Alloys
Metals Reference Book, C. J. Smithells, New York, N.Y., Plenum Press (E15).
Modern Pearlitic Malleable Castings Handbook, Malleable Research and Development Foundation (B49).
Quality Control Handbook, American Society for Quality Control (B17), New York, N.Y., McGraw-Hill.
Rare Metals Handbook, C. A. Hampel, New York, N.Y., Reinhold.
SAE Handbook, Society of Automotive Engineers (B60).
Space Materials Handbook, Aerospace Materials Information Center (D1).
Steel Castings Handbook, Steel Founders' Society of America (B67).
Steel Wire Handbooks, Wire Association, Inc. (B71).
Tool Engineers Handbook, Society of Manufacturing Engineers (B64).

After the metallurgist has built his basic library of textbooks, reference books, and special publications in his areas of interest, the best way for him to keep up to date is to get on the mailing lists of the publishers themselves. The following list represents the most important and prolific commercial publishers of metallurgical materials in English. The criterion for selection was the number of acquisitions made by the ASM library over a two-year period.

Most publishers will be happy to send catalogs and place the requester on the mailing list for new announcements.

ACADEMIC PRESS INC.
111 Fifth Ave.
New York, N.Y. 10003

ADDISON-WESLEY PUBLISHING
Co., INC.
Reading, Mass. 01867

AMERICAN ELSEVIER PUBLISHING
Co., INC.

52 Vanderbilt Ave.
New York, N.Y. 10017

AMERICAN TECHNICAL SOCIETY
848 E. 58th St.
Chicago, Ill. 60637

BARNES & NOBLE
105 Fifth Ave.
New York, N.Y. 10003

BUTTERWORTH & CO. LTD.
88 Kingsway
London, W.C.2, England

CHAPMAN & HALL LTD.
11 New Fetter Lane
London, E.C.4, England

GORDON & BREACH, SCIENCE
PUBLISHERS, INC.
150 Fifth Ave.
New York, N.Y. 10011

HART PUBLISHING CO., INC.
510 6th Ave.
New York, N.Y. 10011

INDUSTRIAL PRESS INC.—BOOK DIV.
200 Madison Ave.
New York, N.Y. 10016

JOHN WILEY & SONS, INC.
605 Third Ave.
New York, N.Y. 10016

MACDONALD & EVANS LTD.
8 John St.
London, W.C.1, England

THE MACMILLAN CO.
866 Third Ave.
New York, N.Y. 10022

MARCEL DEKKER, INC.
95 Madison Ave.
New York, N.Y. 10016

McGRAW-HILL BOOK CO.
330 W. 42nd St.
New York, N.Y. 10036

MIT PRESS
Cambridge, Mass. 02142

NOYES DATA CORP.
Noyes Bldg.
Park Ridge, N.J. 07656

PERGAMON PRESS, INC.
Maxwell House
Fairview Park
Elmsford, N.Y. 10523
or

PERGAMON PRESS LTD.
Headington Hill Hall
Oxford, England

PHILOSOPHICAL LIBRARY INC.
15 E. 40th St.
New York, N.Y. 10016

PLENUM PUBLISHING CORPORATION
227 W. 17th St.
New York, N.Y. 10011

PRENTICE-HALL, INC.
Englewood Cliffs, N.J. 07632

ST. MARTIN'S PRESS, INC.
175 Fifth Ave.
New York, N.Y. 10010

VAN NOSTRAND REINHOLD CO.
450 W. 33rd St.
New York, N.Y. 10001

WILEY INTERSCIENCE, INC.
(see John Wiley & Sons, Inc.)

There are a number of alerting services for new books. Some are so extensive and generalized they are useful primarily to libraries; however, there are two that might be useful to the individual metallurgist.[3,4] By their very nature, they are always after the fact and may not call attention to a book until a few months after its publication.

TRANSLATIONS

Sputnik, more than a decade ago, spawned not only a tremendous furor over the so-called "information explosion," but

[3] *Technical Book Review Index.* Albert F. Kamper, ed. Special Libraries Association, 235 Park Ave., S., New York, N.Y. 10003. Compiled and edited in the Science and Technology Dept., Carnegie Library of Pittsburgh. Monthly, except July and August. Subscription $15.

[4] *New Technical Books.* Arnold Sadow, ed. The Research Libraries, New York Public Library, Fifth Ave. and 42nd St., New York, N.Y. 10018. Monthly except August and September. Subscription $7.50.

also a tremendous interest in Soviet science and technology, as recounted in Soviet publications. Since very few Americans can read Russian, translations into English received a high priority in the nation's scientific efforts. The result was a large number of cover-to-cover translated journals, many of strong metallurgical interest. (One can hardly say they are a dime a dozen; subscription prices for typical metallurgical journals run around $110 to $150.) Following is a list of the most important titles dealing with metallurgy.

AMERICAN INSTITUTE OF PHYSICS (B12)
335 E. 45th St.
New York, N.Y. 10017

1. *Soviet Physics—JETP*
   (*Zhurnal Eksperimental'noi i Teoreticheskoi Fiziki*)
2. *Soviet Physics—Solid State*
   (*Fizika Tverdogo Tela*)
3. *Soviet Physics—Technical Physics*
   (*Zhurnal Tekhnicheskoi Fiziki*)
4. *Soviet Physics—Crystallography*
   (*Kristallografiya*)
5. *JETP Letters*
   (*Zhurnal Eksperimental'noi i Teoreticheskoi Fiziki Pis'ma*)
6. *Soviet Physics—Semiconductors*
   (*Fizika i Tekhnika Poluprovodnikov*)

BRITISH CAST IRON RESEARCH ASSOCIATION (B22)
Bordesley Hall
Alvechurch, Birmingham, England

1. *Russian Castings Production*
   (*Liteinoe Proizvodstvo*)

CONSULTANTS BUREAU
Plenum Publishing Corp. (E15)
227 W. 17th St.
New York, N.Y. 10011

1. *Metal Science and Heat Treatment*
   (*Metallovedenie i Termicheskaya Obrabotka Metallov*)
2. *Soviet Powder Metallurgy and Metal Ceramics*
   (*Poroshkovaya Metallurgiya*)
3. *Protection of Metals*
   (*Zashchita Metallov*)

4. *Industrial Laboratory*
   (*Zavodskaya Laboratoriya*)
5. *High Temperature*
   (*Teplofizika Vysokikh Temperatur*)
6. *Metallurgist*
   (*Metallurg*)

FARADAY PRESS, INC.
84 Fifth Ave.
New York, N.Y. 10011

1. *Soviet Materials Science*
   (*Fiziko-Khimicheskaya Mekhanika Materialov*)

IRON AND STEEL INSTITUTE (B46)
1, Carlton House Terr.
London, S.W.1, England

1. *Steel in the U.S.S.R.* (*Stal'* and
   Izvestiya VUZ Chernaya Metallurgiya)

PERGAMON PRESS, INC. (E14)
Maxwell House, Fairview Park
Elmsford, N.Y. 10523

1. *Physics of Metals and Metallography*
   (*Fizika Metallov i Metallovedenie*)

PRIMARY SOURCES
11 Bleecker St.
New York, N.Y. 10012

1. *Soviet Journal of Nonferrous Metals*
   (*Tsvetnye Metally*)

SCIENTIFIC INFORMATION CONSULTANTS LTD. (E17)
661 Finchley Rd.
London, N.W.2, England

1. *Russian Metallurgy*
   (*Izvestiya Akademii Nauk SSSR Metally*)
2. *Corrosion Control Abstracts*
   (Corrosion and Anti-corrosion Section of
   Referativnyi Zhurnal)

THE WELDING INSTITUTE (B70)
Abington Hall
Abington, Cambridge, England

1. *Automatic Welding*
   (*Avtomaticheskaya Svarka*)
2. *Welding Production*
   (*Svarochnoe Proizvodstvo*)

Publications of other countries do not fare as well as those of the Soviet Union, and the only other important cover-to-cover translation journal in metallurgy is *Giessereiforschung in English*, translated from the German. Its sponsor is A. S. Morrison, Penlee, Thirlstane Rd., Malvern, Worcestershire, England. However, many international journals publish in two or even three languages, usually English, French, and/or German. A number of important Japanese journals also have English editions. *A Guide to Scientific and Technical Journals in Translation* was published in 1968. It covers all languages and all disciplines.[5]

The National Science Foundation has encouraged various translation programs and has provided a considerable amount of financial support. NSF also coordinates and administers the Special Foreign Currency Science Information Program, which employs foreign translators and pays them in foreign currency owned by the United States. The translations made under this program are generally of material older than, and not competitive with, that produced by translation programs of domestic agencies. (See National Technical Information Service, C10, for publication of lists and translations.)

The Joint Publications Research Service is another arm of NTIS which translates research and development literature on a worldwide basis, with emphasis on USSR and East European translations. (See C3 for descriptions of the JPRS standing order service and abstract journals dealing with materials science and metallurgy.) *Transdex*, a separate guide to these translations, is described in the entry for CCM Information Corporation (E3).

For unpublished translations, the best resource is the U.S. National Translations Center, operated by the John Crerar Library (A7). The Center currently contains 160,000 translations

[5] *A Guide to Scientific and Technical Journals in Translation*. Carl J. Himmelsbach and Grace E. Boyd. Special Libraries Association, 235 Park Ave., S., New York, N.Y. 10003, 1968.

and provides searching tools (the *Translations Register-Index* [semimonthly] and the *Consolidated Index of Translations into English*) for both current accessions and retrospective searching. Similar services are provided by the Translations Section of the National Science Library of Canada (A12).

Henry Brutcher (E9), a metallurgist by education, has been preparing and issuing technical translations for over 40 years. The translations are prepared on speculation and offered for general sale. Because multiple copies are sold, the price is considerably less than the price of a translation made for a single customer.

Serving the field of ferrous metallurgy is the British Iron and Steel Industry Translation Service (BISITS) of the Iron and Steel Institute (B46). It is a clearinghouse somewhat similar to the U.S. National Translations Center, and its lists of new translations are widely circulated.

There are many commercial translating services, and even more freelancers, prepared to translate on order from one or a variety of languages. Associated Technical Services, Inc. (E2), for example, offers services in some 35 languages; it also provides extensive lists of previously prepared translations in various subject categories including a number of categories dealing with metals. ATS also has a current awareness and monitoring service covering Russian literature.

A directory of translating personnel and activities is published by the Special Libraries Association.[6] It gives the names and qualifications of 740 freelance translators and 87 commercial translating firms in the United States and Canada. Five indexes cover subject, language, geography, publications, and depositories.

Another source for general information on translators is the American Translators Association, 80 E. 11th St., New York, N.Y. 10003.

Three libraries, in addition to Crerar, maintain important collections of translations that are available for reference purposes. The Slavic Library at Battelle Memorial Institute (F1) buys everything obtainable from the USSR on the subject of ores. The Carnegie Library of Pittsburgh (A5) has an extensive collection of translations, as does the Engineering Societies Library (A6). The latter also provides translating services at a standard fee.

[6] *Translators and Translations: Services and Sources in Science and Technology.* F. E. Kaiser, ed. 2nd ed., 1965. Special Libraries Association, 235 Park Ave., S., New York, N.Y. 10003.

STANDARDS AND SPECIFICATIONS

The three principal organizations in the United States dealing with standards are the National Bureau of Standards (C6), the American Society for Testing and Materials (B18), and the American National Standards Institute, Inc., 1430 Broadway, New York, N.Y. 10018. Although their activities in the area of information services sometimes appear to overlap, each has different functions.

The Bureau of Standards (NBS) does not primarily write and disseminate standards per se, but rather performs the research and then develops actual standard materials and products. Its two key standards programs are the "Standard Reference Data Program" (C7), which provides critically evaluated quantitative information relating to a property of a definable substance or system, and the "Standard Reference Materials Program," which provides well-characterized materials that can be used to calibrate a measurement system or to produce scientific data that can be readily referred to a common base.

The Bureau also sponsors a number of other projects and activities that generate valuable metallurgical information, as described in C6 and C7.

The American Society for Testing and Materials is unquestionably the basic source of published metallurgical standards and specifications, covering everything from A ("Accelerated Life Tests of Iron-Chromium-Aluminum Alloys for Electrical Heating") to Z ("Zirconium and Zirconium Alloy Sheet, Strip, and Plate for Nuclear Applications") in its current *ASTM Book of Standards*. It does not directly undertake research to develop standards, but relies on the cooperative efforts of thousands of scientists, engineers, and specialists who serve on its numerous committees.

The American National Standards Institute (ANSI) is the American clearinghouse for integrating and coordinating standards activity on the national level. It is a federation of more than 100 trade associations, technical societies, professional groups, and consumer organizations. It also has an elaborate committee structure and currently is responsible for 3,600 published standards. ASTM is one of the many contributing organizations, and all of the standards listed in the ANSI catalog under the categories of "ferrous materials and metallurgy" and "nonferrous materials and metallurgy" are also ASTM standards. ANSI represents the United States in the development of international standards and in the work of the technical committees of the International Standards Organization (ISO) and the International Electrotechnical Commission (IEC).

Standards, particularly specifications, are also issued by a number of organizations working in specialized fields. Government agencies are prolific publishers of specifications—usually contained in handbooks and manuals—and particularly military specifications under the aegis of the Department of Defense. Most of these are mechanically or design oriented, although many of the "Mil Specs" do deal with metals and related materials.

Alloy specifications are naturally of the greatest interest here, and most of these are sponsored by trade associations. The best single source for alloy specifications is probably *Metals Handbook, Vol. 1—Properties and Selection of Metals* (see American Society for Metals, B15). Another ASM publication, the *Metal Progress Databook*, has a section on "Materials in Design Data," which gives compositions, properties, and applications in various combinations for some 28 categories of materials.

Engineering Alloys, by Norman E. Woldman, has been an old standby since it was first published by ASM in 1936. The latest (4th) edition was published by Reinhold in 1962. In the late 1950's, Dr. Woldman also established a subsidiary resource known as *Alloy Digest* (for address see the list of prime journals). Each issue consists of a single sheet giving manufacturer, composition, properties, recommended treatments, and applications of a particular alloy. About eight sheets are issued each month.

The following organizations compile and publish specifications and standards of various kinds. The name of the organization is usually a general guide to the subject matter covered.

Aerospace Materials Information Center (D1)
Alloy Casting Institute Division, Steel Founders' Society of America (B67)
Aluminum Association, The (B2)
American Die Casting Institute (B6)
American Foundrymen's Society (B8)
American Hot Dip Galvanizers Association (B9)
American Institute of Chemical Engineers (B11)
American Iron and Steel Institute
American Metal Stamping Association (B13)
American National Standards Institute, Inc.
American Society for Metals (B15)

American Society for Nondestructive Testing (B16)
American Society for Testing and Materials (B18)
American Society of Mechanical Engineers (B14)
American Vacuum Society (B19)
American Welding Society (B20)
Association of Iron and Steel Engineers (B21)
Copper Development Association, Inc. (B31)
Defense Metals and Ceramics Information Center (D11)
Electronic Properties Information Center (D14)
Gray and Ductile Iron Founders' Society (B35)
Industrial Heating Equipment Association (B36)
Instrument Society of America (B42)
Investment Casting Institute (B45)
Lead Industries Association (B47)
Malleable Research and Development Foundation (B49)
Mechanical Properties Data Center (D19)
Metal Powder Industries Federation (B50)
National Association of Corrosion Engineers (B53)
National Bureau of Standards, Office of Standard Reference Data
    (C7)
Non-Ferrous Founders' Society (B54)
Open Die Forging Institute (B55)
Society of Automotive Engineers (B60)
Society of Manufacturing Engineers (B64)
Steel Castings Research and Trade Association (B66)
Steel Founders' Society of America (B67)
Superconductive Materials Data Center (D26)
Thermodynamic Properties of Metals and Alloys, Lawrence
    Radiation Laboratory (D29)
Thermophysical Properties Research Center (D30)
Welded Steel Tube Institute (B69)
Wire Association, Inc. (B71)

Tools for identifying standards and specifications are:

Standards and Specifications Information Sources, Detroit, Mich.,
    Gale Research Co. (E7)
Department of Defense Index of Specifications and Standards,
    Defense Supply Agency[7]

[7] *Department of Defense Index of Specifications and Standards*. Part 1.
Alphabetical Listing; Part 2. *Numerical Listing.* Available from the Superintendent of Documents, U.S. Government Printing Office, Washington, D.C.
20402.

GEDS Weekly Bulletin, Global Engineering Documentation Services, Inc. (E8)

Office of Engineering Standards Services, National Bureau of Standards (C6)

PATENTS

Searching of patents is a very special skill usually requiring the services of an expert, e.g., a patent attorney. What the individual metallurgist should know about patents as an information resource is indicated, in question-and-answer form, by a few excerpts from the small brochure entitled *The Book Collection and Services of the Linda Hall Library—An Outline Guide* (A9):

1. Does the Linda Hall Library have the U.S. patents?
Answer: We have the Patents Specifications and Drawings from July 1946 to date. Patents issued from 1946 to 1958 have been microfilmed . . .
2. Can I make a full patent search in the Linda Hall Library?
Answer: No. The only place where a full patent search is possible is in the U.S. Patent Office.
3. Will the Patent Office make a patent search for me?
Answer: No . . .
4. Will the Patent Office let me know what other patents are issued on a subject, or on a device similar to mine?
Answer: The Patent Office will, without charge, answer requests concerning the proper "Field of Search," that is, list those classes and subclasses which are thought to contain copies of patents pertinent to the subject matter identified in the request . . .
5. How do I get this information?
Answer: By writing a letter to the Commissioner of Patents, Washington, D.C. 20231 . . .
6. What kind of information can I expect to receive from the Patent Office?
Answer: You will receive a statement of the number of patents in the class and subclass into which your patent falls, with a sheet or sheets listing the numbers of all these patents. There is a charge for this service . . .
8. If I follow this procedure—write to the Patent Office, receive a statement of the patent numbers in the related subclasses, and examine the patents themselves—will I have made a patent search?
Answer: No. The only reliable way to make a patent search is to have it done by an accredited patent attorney . . .
9. Is there an index to the U.S. patents available in the Linda Hall Library?
Answer: Yes, there is an *Annual Index to Patents* and an *Index to Trade Marks* . . .
10. What other facilities are available for searching patents in the Linda Hall Library?
Answer: The Patent Office has (in 1960) made available microfilm copies of the official record of the classification of U.S. Patents. This record is a

comprehensive listing of more than 3,000,000 U.S. Patents arranged numerically in Class and Subclass order according to the Patent Office classification. The titles of each class and subclass can be found in the *Manual of Classification* published by the U.S. Patent Office, which is available in the Linda Hall Library . . .

The Library has a complete file of the *Official Gazette of the U.S. Patent Office* and *Annual Index to Patents* . . .

The *Official Gazette of the United States Patent Office* (C17) is the abstracting journal generally used for scanning—if not for authentic patent searching. Its annual subscription price is $78.

It is impossible to obtain metallurgical patents as a group (except through Derwent's as noted below), although there are a number of chemical patent services. (See Chemical Abstracts Service Patent Concordance, F3, and IFI/Plenum Data Corp.'s *Uniterm Index to U.S. Chemical and Chemically Related Patents* and World Chemical Patent Index, E15.)

Worldwide patents are covered by Derwent Publications Ltd. (E4) in its *Central Patents Index*, issued in 12 subject sections with *Alerting Bulletins* for each. To receive *Central Patents Index Alerting Bulletin M—Metallurgy*, one must pay a basic fee of $575 for the *Central Patents Index*, plus $27 for *Bulletin M*. The primary countries covered in each issue are Belgium, Canada, France, West Germany, Japan, the Netherlands, USSR, United Kingdom, and the United States; occasionally, patents of other countries with lesser metallurgical interest will appear. The abstracts are very brief, contain no drawings, and are not nearly of the quality of those in the *Official Gazette* for the United States.

Some of the abstract journals (e.g., *Chemical Abstracts*, *Copper Abstracts*, and *World Aluminum Abstracts*) include patents, while others (e.g., *Metals Abstracts* and *Engineering Index*) do not.

Herner's *Guide* (G2), on page 93 in its Appendix, has a full description of how to obtain copies of patents.

4

FEDERAL AGENCIES—
WHO CAN USE THEM AND HOW

The U.S. Government acknowledged the urgency of the so-called "information explosion" problem quite some time before industry and the scientific community were willing to recognize that it existed. Considerable furor arose among the traditional abstracting and indexing services when Congress (largely spearheaded by Senator Hubert Humphrey) passed the "Science and Technology Act of 1958."

The pros and cons of creating a monolothic government agency for scientific and technical communication were argued for a number of years thereafter. Although it has not materialized, existing agencies—and many new ones—were given the authority to invest very substantial amounts of money and talent in making government-sponsored information available to the general public, particularly to government contractors.

The result is a very loosely knit group of agencies, departments, offices, branches, commissions, and bureaus dealing with scientific and technical information in all its aspects. The overall picture is complicated by the fact that many government agencies interact with private agencies in various ways, including cooperative agreements. Another complication is caused by the fact that the Federal Government is a kaleidoscope of shifting, changing, organizing, reorganizing, or disbanding agencies. These agencies, and their products and services, are identified by a vast collection of acronyms. Moreover, names and acronyms change. Functions of agencies also change over a period of time.

For purposes of examining available information services, government agencies will be considered in the following five categories: those responsible for document distribution; those responsible for generating reports (mission-oriented agencies); those that support information or data centers; those that provide library services; and miscellaneous.

AGENCIES RESPONSIBLE FOR DOCUMENT DISTRIBUTION

There are three major agencies and one minor in this category:

Superintendent of Documents, U.S. Government Printing Office (GPO) (C12)
National Technical Information Service (NTIS) (C10)
Defense Documentation Center (DDC) (C2)
Joint Publications Research Service (JPRS) (C3)

Publications that are offered free of charge usually are not handled by the above agencies but are distributed by the issuing agency. The Superintendent of Documents handles almost everything the government prints and offers for sale, with the exception of the report literature; NTIS is the distributing agency for government-generated science and technology reports in the public domain; and DDC is primarily, although not exclusively, the source for Department of Defense (DOD) classified or "limited distribution" documents. Agencies other than DOD, such as AEC and NASA, handle the dissemination of their own classified documents. JPRS, a branch of NTIS, is concerned solely with providing translations of foreign materials.

Prices of all government publications are modest indeed and hardly on a cost-recovery basis. In recent years, NTIS vastly simplified the purchase of government reports by setting standard unit prices: hard copy—$3.00 for 1–300 pages, $6.00 for 301–600 pages, $9.00 for 601–900 pages; microfiche—$0.95 per document. Another recent innovation is Selected Categories in Microfiche (SCIM) offered by NTIS (C10), which is an inexpensive way to obtain substantial quantities of government reports in a variety of categories.

All four of the above distributors provide regularly published alerting services. The *Monthly Catalog* of the U.S. Government Printing Office covers everything published by the government and available from GPO, including the huge stockpiles of congressional deliberations. It is primarily a library tool and is hardly recommended for searching the metallurgical literature on specific topics. *Government Reports Announcements* (GRA), together with its *Index* and the subsidiary "finding" resources of NTIS, and the *Technical Abstract Bulletin* (TAB) of DDC do lend themselves to metallurgical searching of the report literature. Subscription prices are reasonable: $30 for *GRA*, $22 for its *Index*; *TAB* is free, but only to "registered" DDC users. The charge for the JPRS abstract services in materials science and metallurgy is modest.

MISSION-ORIENTED AGENCIES

"Mission-oriented" is the documentalist's term for organizations that direct their attention toward a particular end product or application area that is essentially nondisciplinary in nature. Most government agencies that offer information services in particular fields are of this type, whereas most of the scientific and technological disciplines are covered by the technical societies, trade associations, and other private organizations. This divides the responsibility neatly and appears to be working out well.

Most of the mission-oriented agencies are actively engaged in research and development, with information services as a sideline —albeit a large and an important one. The major government agencies for metallurgical information are:

Department of Defense (DOD)
National Aeronautics and Space Administration (NASA) (C5)
National Bureau of Standards (NBS) (C6 and C7)
U.S. Atomic Energy Commission (AEC) (C15)
U.S. Bureau of Mines (C16)

DOD, NASA, and AEC are responsible for a vast quantity of reports. (According to the ASM definition of this discipline, it is estimated that annually about 5,000 unclassified reports are heavily concerned with metallurgy.) In addition to the above agencies, reports are generated by other government departments engaged in research and development, namely, the Army, Navy, and Air Force. Government contractors, including industry, universities, and research institutes, also issue reports. (The U.S. Bureau of Mines might be construed as being disciplinary, but its interests far transcend the usual concept of mining as a technological discipline.)

The alerting services for government reports of NTIS and DDC have already been mentioned. To them should be added NASA's *Scientific and Technical Aerospace Reports* (STAR), annual subscription $54, *Index* $30; and AEC's *Nuclear Science Abstracts* (NSA), annual subscription $42, *Index* $38.

The other information services of these agencies, including the National Bureau of Standards and the U.S. Bureau of Mines, are varied and divergent. Consult the individual descriptions of each in the directory section.

SPONSORS OF INFORMATION OR DATA CENTERS

The Department of Defense (Defense Supply Agency), National Aeronautics and Space Administration, U.S. Atomic Energy Commission, and National Bureau of Standards each sponsors or funds (sometimes jointly) a number of information centers of metallurgical interest. These are described in the following chapter. The Air Force, Army, and Navy also sponsor information centers, mostly subsidiary to a research and development program at a government laboratory.

The *Directory of Federally Supported Information Analysis Centers* (G4), issued by COSATI, lists 119 centers, and the *Directory of the Defense Documentation Center Referral Data Bank* (G3) lists 182; of these, 21 are of metallurgical interest and are covered both in Chapter 5 and in the directory section. These centers cover a very wide range of subject matter as well as types of services offered.

AGENCIES WITH LIBRARY SERVICES

Conventional library services of government agencies are surprisingly limited, although most profess to be open to the public for reference purposes. Loans and related services are generally limited to government personnel and government contractors, although the line of demarcation between contractors and the general public is blurred by the numerous subcontractors and suppliers who might have a legitimate "need to know."

The Library of Congress (A8) is the major government library. Since numerous depository libraries for all types of government publications have been designated throughout the country, individual agency libraries are hardly necessary for materials in the public domain. The *Directory of Special Libraries and Information Centers* (E7 and G7) by Kruzas contains a list of more than 800 depository libraries for government publications, as well as lists of libraries with United States patent files and libraries holding AEC reports.

MISCELLANEOUS

Before closing this chapter on government information services, mention should be made of the Committee on Scientific and Technical Information (COSATI) of the Federal Council for Science and Technology and the National Science Foundation (NSF), which play special roles in the information field.

COSATI was formed in 1962 in recognition of the need for a government-wide organization to cope with problems of scientific and technical information. It deals with neither disciplines nor missions per se, but has a strong influence on the directions taken by the various agencies that supply services to the public. (For a description of the *COSATI Subject Category List,* see NTIS, C10, under "Self Helps.")

The National Science Foundation exercises an even greater influence on the direction of information activities by providing financial support. The Office of Science Information Service (OSIS) of NSF provides support primarily to private organizations developing and operating information services and systems, but government agencies can also be recipients. Many of the abstracting and indexing services, as well as the more elaborate information retrieval services, currently receive financial support or have received it in the past. However, most grants are made to support research and development and not to underwrite operating costs.

In addition, NSF partially supports the Science Information Exchange of the Smithsonian Institution (C11) and coordinates and administers the Special Foreign Currency Science Information Program (see *SFCSI List of Translations in Process,* National Technical Information Service, C10).

With the curtailment of federal support for science and technology toward the end of the 1960's, less money has been made available for scientific and technical information services. Many services that for years had provided materials to government personnel and contractors on request are now required to implement service charges. For example, in mid-1971, the Department of Defense directed that all of the information analysis centers that it sponsors should institute a system of charges for their products. These fees are generally modest in comparison to those that have been found necessary to cover the cost of services provided by nongovernmental agencies.

5

SEARCHING SERVICES AND
INFORMATION CENTERS

Literature searching has been performed by librarians, as well as by scholars and research workers, for centuries. However, it is only in the last twenty years or so that data processing and related equipment has been applied to the task. The use of automatic and semiautomatic procedures for literature searching is now generally referred to as "information retrieval."

Basically, there are two types of searching services: *retrospective search* and *current awareness.* The latter involves procedures designed to keep the practitioner up-to-date with currently issued materials; the former involves searching of the past literature in response to specific requests. There are at least five types of purveyors of retrieval services: libraries, information dissemination centers, information analysis centers, data centers, and organizations that provide services as an adjunct to other activities (principally technical societies and some commercial organizations).

When a user goes to a surrogate for information, he is not likely to be overly concerned about the retrieval method used so long as he gets the desired results, at a price he can afford, within the required time limits. However, the following generalities should assist the user in determining what source he wants to consult.

METHODS

Manual methods consist of visually searching library card catalogs, printed indexes, abstracts, and bibliographies. When charges are made for this type of service, they are usually calculated on an hourly basis. Delivery time and cost are influenced by the extent and complexity of the search and the size of the output. When a metallurgist decides to consult an organization that utilizes manual methods—usually a library—he might wish to match the abstracting and indexing services appropriate to his topic (as listed in Chapter 2) against the particular library's holdings, and thus get an idea of the type of resources that can be searched.

Semiautomatic methods of information retrieval include the use of marginal punched cards, inverted term cards, and "peek-a-boo" or optical coincidence cards. These methods are used primarily for organizing personal files, although some librarians and information centers may use them for relatively small, specialized collections. Cost and delivery time are influenced by the same factors that affect manual searching.

Computer methods are now widely used and are commonly employed in various types of information centers. Most machine-based retrieval systems still function in the "off-line" mode. For example, a request for a search is made to an information center. This request is translated into a searching strategy and then matched against the machine data base by computer. Results are usually mailed to the requester in the form of a computer print-out or similar hard copy (reproductions of printed abstracts, for example). More recent retrieval systems are operated in an on-line mode. Using a typewriter or video (cathode ray tube) terminal connected to a computer by some type of communication line, the searcher is able to directly interrogate a machine-readable file of document records. At the present time, only a few such systems that have metallurgical information are operative, and only a few users have access to the remote consoles. One of the best examples is NASA/RECON (see C5). AEC is also using the RECON system on an experimental basis (see Lawrence Radiation Laboratory, D17). A similar system is Battelle's BASIS-70, now being used by the Copper Data Center (D8). Overall cost of operation tends to be high, but the time-sharing principle (which is practical only for large, heavily used files), together with a possible reduction in rates for telecommunication, should bring about a much wider use of this method in the future. On-line retrieval allows, theoretically at least, virtually immediate response.

For extensive or complex searches of large files, computer methods generally are faster and cheaper than manual methods. Other advantages are that costs can be fixed and delivery times are accurately predictable. To get the best possible results from a machine-based system, it is desirable that the user be familiar with how the information is organized in the file and with the rudiments of how the file is searched. He can then phrase his questions properly and will have some idea of what he is likely to receive. The "search analyst" on the staff of an information center should be thoroughly familiar with the contents of the various computer files. Once the user has described his need, the analyst can predict with some confidence the likelihood of retrieving the desired information and can give the user an approximate estimate of the amount of relevant literature likely to be retrieved. The metallurgist who uses an on-line system directly must, of course, be much more knowledgeable of the file content and the method of querying it than one who delegates his search to a librarian or other information specialist.

TYPES OF SERVICE

The user may choose to use one source for retrospective search and another source for current awareness. It is simpler to use only one source and, thus, avoid the necessity of rephrasing the question for each system. Manual methods are more likely to be used for searches that go back a decade or more, because few computer-based files are more than five years old, and many of them offer current awareness service only. There are some who contend that, for retrospective searching, manual methods are almost invariably better and cheaper than computer methods, whereas the computer provides current awareness or alerting services more efficiently and economically. This, however, is a dangerous generalization. Certain types of retrospective search, involving complex conceptual relationships, are virtually impossible to handle with anything except a computer system. In doing a retrospective search off-line by computer, the question must be properly phrased at the start, as it is not possible to change direction in midstream. However, in manual searching (and in on-line systems), related concepts and indexing terms often suggest themselves during the course of the search.

A special type of current awareness or alerting service is Selective Dissemination of Information (SDI). The recipient of such service describes his current professional interests, and these are translated into a formal "interest profile" in the language of the system. The profiles are stored in machine-readable form and are matched, on a regular basis (weekly or monthly), against records of documents newly added to the file. When a document matches a profile, it is brought to the attention of the person for whom this profile was constructed.

Instead of—or in addition to—current awareness services tailored to the individual's interest profile, the supplier may offer "group profiles" corresponding to various subtopics covered by the total data base. These are sometimes referred to as "standard interest profiles" (as opposed to "custom interest profiles"). The cost of subscribing to a standard interest profile usually is considerably less than the cost of subscribing to the more personalized services because the search results of a single computer run can be mass produced for a number of customers. Descriptions in directory section D generally indicate which of these services are offered.

In deciding which of several services (e.g., information centers) to consult for information on a particular topic, it may be helpful to consult the National Referral Center (C8) at the Library of Congress. This Center is a clearinghouse on where to go for such information.

The following is a list of organizations that produce magnetic tape files (data bases) of bibliographic records. It is arranged in three categories: organizations that "manufacture" the data base only (often a computerized index to a printed abstract journal) and then lease or license the files to an information center for searching; organizations that produce a data base and search it exclusively; and organizations producing data bases that can be searched either by themselves or by a licensed information center or agency. Full descriptions of commercially available tape services in many disciplines are contained in the *Survey of Scientific-Technical Tape Services* (G18) by Kenneth D. Carroll. *A Guide to a Selection of Computer-Based Science and Technology Reference Services in the U.S.A.* (G12), compiled by the American Library Association, contains information on machine-readable sources produced by 18 agencies representing professional societies, government, and private organizations.

1. Producers Only (Base Leased to Others, No Retrieval Service)

American Institute of Physics (B12), SPIN (Searchable Physics Information Notices).

CCM Information Corporation (E3), PANDEX Current Index to Scientific and Technical Literature.

Chemical Abstracts Service (F3) *CA* Condensates, *Chemical Titles,* and Patent Concordance.

Engineering Index, Inc. (F4), COMPENDEX.

2. Producers With Exclusive Retrieval Service

Alloy Data Center (D4), *Permuted Materials Index* and *Author Index.*

Copper Data Center (D8), Technical Files.

Cryogenic Data Center (D9), Data File.

Defense Metals and Ceramics Information Center (D11), Technical Files.

Electronic Properties Information Center (D14), Literature Files.

Liquid Metals Information Center (D18), Literature File.

Machinability Data Center (D3), Data File.

Mechanical Properties Data Center (D19), Data File.

Science Information Exchange (C11), Registry of Research in Progress.

Thermophysical Properties Research Center (D30), Literature File.

University Microfilms (E20), DATRIX.

3. Producers With Retrieval Service and Leasing Arrangements

American Society for Metals (B15), *Metals Abstracts* META-DEX.

Defense Documentation Center (C2), *Technical Abstract Bulletin.*

IFI/Plenum Data Corp. (E15), *Uniterm Index to U.S. Chemical and Chemically Related Patents* and World Chemical Patent Index.

Institute for Scientific Information (E10), Source Files and Citation Files.

Institution of Electrical Engineers (B40), INSPEC.

National Aeronautics and Space Administration (C5), *Scientific and Technical Aerospace Reports* and *International Aerospace Abstracts.*

National Technical Information Service (C10), *Government Reports Announcements.*

U.S. Atomic Energy Commission (C15), Indexes to *Nuclear Science Abstracts.*

PURVEYORS OF RETRIEVAL SERVICES

Libraries: In general, special libraries are a more useful source of metals information than public libraries. The user associated with an industrial company or university that maintains a good library, with qualified librarians, is fortunate; although, as noted in Chapter 1, not enough users take full advantage of available library facilities. The following libraries have good collections in the metals field and provide literature searches at hourly rates:

John Crerar Library (A7)
Engineering Societies Library (A6)
Science and Technology Division of the Library of Congress (A8)

Information Centers: The rise of the information center is a modern-day phenomenon stimulated by the science- and information-conscious society of the '60s and the rise of computer information-processing systems and techniques which can be operated only by highly skilled and highly trained personnel. There are three general types of information centers, although, in practice, some of the services overlap: information dissemination centers, information analysis centers, and data centers.

The *information dissemination center,* sometimes called a "document retrieval service," has been characterized as a middleman or broker that operates between the manufacturer or producer of the data base and the user. By providing specialized searching services, such a center performs what is sometimes known as a "repackaging" function—selecting from one or more indexes or from abstract files only those items of published information that relate to a particular subject area, as defined by a set of query specifications. (Searches can usually be qualified by author, source, date, subject, and other characteristics of the document.) They usually (but not necessarily exclusively) rely upon computerized data bases. Information dissemination centers neither interpret results of a search nor provide critical analysis. Some concentrate their efforts on current awareness services; others provide retrospective search only or a combination of current awareness and retrospective search services.

Perhaps the best known information dissemination centers are the seven NASA "Regional Dissemination Centers," sponsored by the NASA Office of Technology Utilization (C5) in an effort to make the results of government aerospace research available to industry. Most of these centers use a number of other data bases in addition to the NASA files, as can be seen from their descriptions in the directory section.

An excellent description of the functions and services of a typical information dissemination center is contained in a 22-page brochure, with the improbable title of *Kascapabilities,* issued by the Knowledge Availability Systems Center (D16) at the University of Pittsburgh. It contains quite specific information concerning the content, number of items, and types of indexing used in the various data bases to which the Center subscribes, including the indexes to *Scientific and Technical Aerospace Reports* (C5) and *International Aerospace Abstracts* (B10), the NTIS *Government Reports Announcements* (C10), *Metals Abstracts* META-DEX (B15), *Engineering Index* COMPENDEX (F4), and *CA* Condensates and *Chemical Titles* (F3).

A number of independent dissemination centers, covering multiple data bases, now provide machine-searching services to individuals or institutions on a subscription basis. One such center exists at IITRI (Illinois Institute of Technology Research Institute, D15) and another at the University of Georgia, D32.

Following is a list of major information dissemination centers which offer metallurgical searches. NASA Regional Dissemination Centers are indicated by an asterisk.

Aerospace Research Applications Center (D2)*
Dow Current Awareness Service (D13)
IIT Research Institute (D15)
Knowledge Availability Systems Center, University of Pittsburgh (D16)*
Lawrence Radiation Laboratory (D17)
Mechanized Information Center, Ohio State University (D20)
New England Research Application Center (D21)*
North Carolina Science and Technology Research Center (D23)*
Technology Application Center, University of New Mexico (D27)*
Technology Use Studies Center, Southeastern State College (D28)*
University of Calgary Data Centre (D31)
University of Georgia Computer Center (D32)
Western Research Application Center (D34)*

Information analysis centers are defined by Panel 6 of COSATI (G4) as follows:

An information analysis center is a formally structured organizational unit specifically (but not necessarily exclusively) established for the purpose of acquiring, selecting, storing, retrieving, evaluating, analyzing, and synthesizing a body of information and/or data in a clearly defined specialized field or pertaining to a specific mission with the intent of compiling, digesting, repackaging, or otherwise organizing and presenting pertinent information and/or data in a form most authoritative, timely, and useful to a society of peers and management.

In other words, such a center provides critical evaluation of retrieved literature references and the information contained therein. It usually provides consultation services as well.

Typical information analysis centers are sponsored by the Department of Defense and the Atomic Energy Commission. Perhaps one of the oldest and best known as far as metallurgical information is concerned is the Defense Metals and Ceramics Information Center (MCIC) (D11) at Battelle Memorial Institute. Originally, the services of such centers were available at no charge, but only to "qualified personnel" (government agencies, contractors, subcontractors, and suppliers). Now, in most instances, anyone can use their services for a modest fee. National Technical Information Service (NTIS) (C10) serves as the principal marketing agent for publications and information products of nine DOD-sponsored information analysis centers. Of the nine, five are of metallurgical interest; they are marked on the list by an asterisk. The Army also sponsors centers of this type, but generally these are operated solely for the personnel of the Army R&D effort involved. Following is a list of information analysis centers covering metallurgical interests:

Army Materials and Mechanics Research Center (D5)
CIDEC Information Service (B30)
Cobalt Information Center (D7)
Copper Data Center (D8)
Defense Metals and Ceramics Information Center (D11)*
Electronic Properties Information Center (D14)*
Liquid Metals Information Center (D18)
Machinability Data Center (D3)*
Mechanical Properties Data Center (D19)*
Nondestructive Testing Information Analysis Center (D22)
Rare-Earth Information Center (D24)
Research Materials Information Center (D25)
Thermophysical Properties Research Center (D30)*
Watervliet Arsenal Benet Laboratories (D33)

Data centers provide actual data—such as property data—assembled and evaluated at the center rather than references only. A data center is discipline-oriented rather than mission-oriented and, according to the COSATI Ad Hoc Group on Data Centers,[1] should have at least the following capabilities: an information system about the data at the center and the availability of specialized data collections at other locations; a specialized technical library and an automated document retrieval system; a professional staff that carries on data analysis and synthesis; and a professional staff in computer and information sciences.

Typical data centers are those supported by the National Bureau of Standards through its Office of Standard Reference Data (C7) and other divisions. Many information analysis centers also qualify as data centers because of the consulting and advisory services they provide; thus, the following list of data centers contains some organizations that fall in both categories:

Alloy Data Center (D4)
Binary Metal and Metalloid Constitution Data Center (D6)
Cryogenic Data Center (D9)
Crystal-Data Center (D10)
Diffusion in Metals Data Center (D12)
Electronic Properties Information Center (D14)
Machinability Data Center (D3)
Mechanical Properties Data Center (D19)
Research Materials Information Center (D25)
Superconductive Materials Data Center (D26)
Thermodynamic Properties of Metals and Alloys (D29)
Thermophysical Properties Research Center (D30)

As noted in the preceding chapter, several government agencies sponsor or support all three kinds of information centers.

The COSATI *Directory of Federally Supported Information Analysis Centers* (G4) lists 119 centers of the analysis and data types, but does not include dissemination centers.

The three types of centers described above are intermixed in

[1] Vette, J. I. *Report of Ad Hoc Group on Data Centers.* Committee on Scientific and Technical Information (COSATI) of the Federal Council for Science and Technology. Springfield, Va., National Technical Information Service, September 1969. PB 195 523.

directory section D. Individual descriptions will indicate whether current awareness, retrospective search, or both types of services are offered. User qualifications are also given.

Organizations that Provide Information Retrieval Services as an Adjunct to Other Activities (other than libraries and information centers). Such services are provided primarily by technical societies, some commercial companies, and a few nonprofit organizations. Most produce their own data bases in the form of abstracts and indexes, either computerized or printed, although a few rely upon purchased or leased material. Those that produce and search computerized data bases are shown in categories 2 and 3 on page 51. Others are:

American Foundrymen's Society (B8). Document Retrieval, Current Awareness, and Special Literature Search Services.

American Society for Metals (B15). METALERT and Retrospective Information Retrieval Service.

Battelle Memorial Institute (F1). Literature Searches and State-of-the-Art Surveys.

Defense Documentation Center (C2). Searches of the Report Literature and of Ongoing Research Projects.

Franklin Institute, Science Information Services (F5). Literature Searches and State-of-the-Art Surveys.

IFI/Plenum Data Corp. (E15). Chemical Patent Service Bureau.

Institute for Scientific Information (E10). ASCA (Automatic Subject Citation Alert) and ISI Search Service.

Iron and Steel Institute (B46). ABTICS (Abstract and Book Title Index Card Service) and ISIP (Iron and Steel Industry Profiles).

National Aeronautics and Space Administration (C5). SCAN (Selected Current Aerospace Notices).

National Bureau of Standards, Metallurgy Division and Inorganic Materials Division (C6). Literature Searching, Office of Engineering Standards Services, and Standards Searching.

National Bureau of Standards, Office of Standard Reference Data (C7). Inquiry Referral Service.

National Research Council of Canada (C9). Checklists, Digests, and Tech Briefs.

National Science Library of Canada (A12). CAN/SDI Program for Computerized Current Awareness Service and a Question and Answer Service.

National Technical Information Service (C10). GRTA (*Government Reports Topical Announcements*), FAS (*Fast Announcement Service*), NTISearch, and SCIM (Selected Categories in Microfiche).

Production Engineering Research Association (B56). Express Monitoring Service and Literature, Patent and Product Surveys.

Science Information Exchange (C11). Searches of Ongoing Research.

Tin Research Institute (B68). Literature Surveys.

3i Company (E19). Retrospective and Current Awareness Searches of *Engineering Index* COMPENDEX, *CA* Condensates, and INSPEC tapes.

University Microfilms (E20). DATRIX, Retrieval Service for Dissertations.

U.S. Atomic Energy Commission, Division of Technical Information (C15). Reference and Searching Services based on *Nuclear Science Abstracts*, both manual and computerized.

Information centers are experiencing a period of rapid development and transition (for instance, from free services for government personnel and contractors to paid services for industry). How successful they will be in the long run, and what form they will eventually take, will depend largely upon standardization of computer format and search programs.

6

SELF-HELPS—WHAT TO DO WHEN YOU WANT TO ESTABLISH YOUR OWN SYSTEM

Every metallurgist or engineer has a personal technical library even though it may consist of only a few books or a few file folders. The methods used to control the personal library range all the way from relying on one's memory to setting up quite elaborate filing systems, many incorporating mechanical devices, and a few even requiring computerized systems.

The methods of indexing and filing are many and varied, but the scientist or engineer should beware of the pitfalls of becoming so intrigued with the gadgetry of a system that he finds himself spending excessive time putting material in but never needing to get it out—or that it's too much trouble to get it out. (Mooers' Law, which the metallurgist may not be familiar with, states that "An information retrieval system will tend not to be used whenever it is more painful and troublesome for a customer [user] to have information than for him not to have it.") Still, the elaborate system may more than compensate for its trouble on some critical occasion when the metallurgist is able to retrieve the precise information he needs at the right time. In all systems, one must be resigned to the fact that some time will be wasted indexing and filing information that never will be used again.

The intent of this book is not to deal with the pros and cons of various indexing methods nor with the devices used to manipulate files, but rather to deal with the various publications and index guides that are concerned specifically with metallurgy or some aspects of it. We are concerned here with the intellectual content of the system rather than how to manage it. Systems and devices are adequately covered in the Herner *Guide* (G2) as well as in other texts, notably Jahoda[1] and Foskett.[2] Another reference on systems and devices is particularly apropos because the author, Dr. Campbell,[3] is information officer of the Pressed Steel Company Ltd. and is, therefore, metallurgically oriented.

Other metallurgical information resources that will be helpful are the SLA *Guide to Metallurgical Information* (G11), the British *How to Find Out in Iron and Steel* (G13), the Crerar Library's *Guide to Metals Literature* (A7), and the *Guide to Literature on Metals and Metallurgical Engineering* (G10) issued by the American Society for Engineering Education.

Two publications of related interest are the ASTM *Manual on Methods for Retrieving and Correlating Technical Data* (B18) and *Modern Techniques in Chemical Information* in preparation by IIT Research Institute (D15); they contain many ideas that will serve the metallurgist as well as the chemist.

The basics of the personal library are the books, journals, and reference publications treated in Chapter 3. Abstract journals and indexes would be excellent additions, but most of the abstract journals are now so costly that they can only be consulted in a company or public library. Some of the searching services covered in Chapter 5 can also be very helpful in building personal files of abstracts on specialized topics.

Books and reference publications include both metallurgical glossaries and foreign-language dictionaries. The best of the former are those by Merriman,[4] Birchon,[5] and "Definitions Relating to Metals and Metalworking" in *Metals Handbook, Vol. 1* (B15). A number of commercial publishers that specialize in foreign-language dictionaries include the following:

[1] Jahoda, G. *Information Storage and Retrieval Systems for Individual Researchers.* New York, Wiley Interscience Division of John Wiley & Sons, 1970.

[2] Foskett, A. C. *A Guide to Personal Indexes.* 2nd ed. Hamden, Conn., The Shoe String Press, Inc., 1967.

[3] Campbell, D. J. "Making Your Own Indexing System in Science and Technology (Classification and Keyword Systems)." *Aslib Proceedings,* 15(10), October 1963: 282–303.

[4] Merriman, A. D. *A Concise Encyclopedia of Metallurgy.* New York, N.Y., American Elsevier Publishing Co., 1965.

[5] Birchon, D. *Dictionary of Metallurgy.* New York, N.Y., Philosophical Library, Inc., 1965.

Adler's Foreign Books, Inc. (E1)
American Elsevier Publishing Co.[6]
Associated Technical Services, Inc. (E2)
Gale Research Co. (E7)
Pergamon Press, Inc. (E14)
Stechert-Hafner, Inc. (E18)

The Index: There are no set rules to determine at what point an index is required for a personal library. Books and reference compilations generally carry their own indexes, and usually the user is familiar enough with his own books to have a pretty good idea of what they contain. But it is when his files become loaded with separates, photocopies, reprints and preprints, technical correspondence, manufacturers' literature, even microfiche—and when he cannot find the particular item of information he knows is buried therein—that the time spent in indexing will pay off.

Setting up an index is a highly specialized and highly individualized art, and it is natural for an information scientist to feel that no one but a specialist can do it properly. This is undoubtedly true for large collections; but, for the small collections in a personal library, it can be done quite adequately if the metallurgist organizes his plan properly and does not deviate from it. An "authority list" or "controlled vocabulary" will help to avoid inconsistencies and to ensure that search patterns take the right directions through the *see* and *see also* references. Controlled vocabularies may take the form of classification schemes, thesauri, subject heading authority lists, keyword lists, and variations thereof. A number of these vocabularies are available to the metallurgist, as follows:

The Aluminum Association, *Thesaurus of Aluminum Technology* (B2).

American Institute of Chemical Engineers, *Chemical Engineering Thesaurus* (B11).

American Society for Metals, *ASM Thesaurus of Metallurgical Terms* and *ASM-SLA Metallurgical Literature Classification* (B15).

Copper Development Association, Inc., *Thesaurus of Terms on Copper Technology* (B31 and D8).

Defense Documentation Center, DOD *Thesaurus of Engineering*

---

[6] *Elsevier's Dictionary of Metallurgy.* In six languages. Compiled by W. E. Clason. New York, N.Y., American Elsevier Publishing Co., 1967.

and Scientific Terms (C2) ; also available from Engineers Joint
Council, 345 E. 45th St., New York, N.Y. 10017.

Engineering Index, Inc., *Engineering Index Thesaurus* and *Subject Headings for Engineering* (F4).

Iron and Steel Institute (London), *Universal Decimal Classification—Special Subject Edition for Metallurgy* and *Simplified
UDC Schedule—Occasional Bulletin No. 3* (B46).

National Aeronautics and Space Administration, *NASA Thesaurus* (C5).

National Association of Corrosion Engineers, *Abstract Filing
System Classification* (B53).

National Technical Information Service, *COSATI Subject Category List* (C10).

U.S. Atomic Energy Commission Division of Technical Information, *Subject Headings Used by the USAEC Division of Technical Information* and *INIS Thesaurus* (C15).

The Welding Institute, *Multilingual Collection of Terms for
Welding and Allied Processes* (B70).

A thesaurus is one of the simplest and most effective types of
vocabulary control. Most thesauri are organized using similar
principles, but the terms included—and the cross-references—do
not always coincide with one another. It is therefore desirable to
use the thesaurus that most closely covers your field of interest.
There is nothing wrong with adding terms peculiar to one's special interests, but caution must be exercised not to violate the-
saurus-building rules. It would be helpful for those who index
fairly generous amounts of abstracts from a single source to
adopt the vocabulary control used by that abstracting service. For
example, a file which contains abstracts from *Metals Abstracts*
should probably be built around the *ASM Thesaurus;* for those in
the steel industry that rely on the British ABTICS and/or Iron
and Steel Industry Profiles service, the *Universal Decimal Classification* (Abridged Edition) might be the best authority. The
COSATI Subject Category List is used by many government
agencies; but it is extremely broad in its interests, therefore,
metallurgical topics are somewhat fragmented. However, an individual with wide subject interests and a large amount of government report literature might find it helpful as a basic or
general outline.

The above is a list of major vocabularies only; many others
have been developed by various organizations and individuals for

specialized subject collections. The source for these is the Bibliographic Systems Center at the School of Library Science, Case Western Reserve University, Cleveland, Ohio 44106. This Center administers a special collection of classification schemes and subject heading lists in almost every field. For $8, the Center will make a computer search of its holdings and send descriptive "clips" of classifications, lists, and thesauri on the desired topic. A recent search brought forth about 35 such schemes on metallurgy (which included corrosion and other related topics). The Center makes no claim as to quality of the items, and many are quite old. An advantage is that copies of selected classifications are available, on loan, for a period of one month. The Special Libraries Association, which sponsored the collection for four decades up to 1966, has published a bibliography of the holdings entitled *Selected Materials in Classification.*[7]

Also, the Engineering Societies Library (A6) publishes a *Bibliography on Filing, Classification and Indexing Systems, and Thesauri for Engineering Offices and Libraries.*

Devices for Retrieval: For the individual who prefers something more sophisticated than the standard card catalog, the American Society for Metals many years ago developed an edge-notched, punched-card filing system designed to be used with the *ASM-SLA Metallurgical Literature Classification* (B15). The latest edition of the *Classification* was published in 1958, but the system is still useful because the user can easily update and expand it. The punched cards are still being manufactured and may be purchased from E-Z Sort Systems Parkison Agency, 35 E. Wacker Drive, Chicago, Ill. 60601.

During that same epoch, similar semiautomatic systems were designed for various related fields. One variation is the optical coincidence card system, which is quite popular and easy to use. A number of types and styles are described by Jahoda.[1]

The use of computers to assist an individual with his personal library could itself be the subject of a book. Not many metallurgists would have the time or the inclination to become deeply involved in a computer retrieval system, although methods of man-machine interaction are becoming increasingly simple. The use of computers for information retrieval services in large com-

---

[7] *Selected Materials in Classification—a Bibliography.* Compiled by B. Denison. New York, N.Y., Special Libraries Association, 1968. $10.75.

panies is rising. An example is the Babcock & Wilcox Research Center in Alliance, Ohio, which maintains a company-wide computerized information system using data bases leased from a number of the producers listed on page 51. The user of the system must absorb a fairly detailed set of instructions for "profile" development. As noted in the chapter on retrieval services, the time will come when a metallurgist can have a remote console on his desk, connected to a computer located in a library or other information center. In such a situation, instantaneous video displays are also possible as a substitute for, or an adjunct to, the typewriter printout.

Collecting Documents: When the metallurgist has his library and filing system organized and in working order, he will undoubtedly start gathering documents with great enthusiasm. This can become quite frustrating, especially when he is confronted with a particularly tempting abstract but has no idea where to find (or look for) the full paper. The rise of the microfilmer and photocopier has considerably eased the procurement of documents despite some outcries from copyright owners. Many of the abstract services provide document backup. For example, the American Society for Metals (B15) has a photocopy service that will make copies of about 80 percent of the papers abstracted in both *Metals Abstracts* and *World Aluminum Abstracts*. Almost all libraries provide photocopy services.

Two unique services are offered by the Institute for Scientific Information (E10): "Original Article Tear Sheet" service (OATS), based on 4,000 source journals; and "Request-a-Print" cards for requesting copies from authors.

Where and how to get government reports, patents, translations, and other specialized types of documents are covered in other chapters.

Other Helps: Home study courses and training manuals are worth mentioning, although they are covered only casually in the directory descriptions. Many of the technical societies and trade associations provide this type of service, even though it is not always mentioned in the individual descriptions. Examples are the Metals Engineering Institute of the American Society for Metals (B15) and the Programmed Learning Courses of the Society of Manufacturing Engineers (B64).

For the metallurgist who really wants to do a professional job

of organizing his resources or wants to set up a company library
—and has the money to spend—there are a number of consultants
and consulting services available. The Special Libraries Association, 235 Park Ave., S., New York, N.Y. 10003, has a number of
helpful publications and suggestions on this subject.

7

WHO IS DOING WHAT IN RESEARCH

I t is often extremely important to find out what is being done in research before the results reach the printed page. Many scientists and engineers claim that the best (and often only) way to obtain up-to-the-minute information is to pick up the phone and call someone you know who is working in the field; in other words, to use the informal communication channels of the "invisible college." Alternatively, there are a few alerting services, mainly government-sponsored, and a few published resources that deal with ongoing research which can refer you to sources outside your own circle of personal acquaintances.

Government Agencies: A starting point for obtaining broad and general guidelines on where to go for information on ongoing research is the National Referral Center at the Library of Congress (C8). The Science Information Exchange (C11)—*the* national registry of government-sponsored research in progress— has excellent metallurgical coverage and offers an alerting service tailored to individual specifications. A volume describing 4,000 research projects in the field of materials has been published. The Defense Documentation Center's Research and Technology Work Unit Information System (C2) also provides an alerting service to user specifications.

Annual research reviews are published by NASA (C5)— known as the *NASA Research and Technology Operating Plan Summaries*—and by the U.S. Atomic Energy Commission (C15) —known as *Research Contracts in the Physical Sciences*. Two government-sponsored information centers also provide research reviews: Defense Metals and Ceramics Information Center (D11) publishes *Reviews of Recent Developments* (variable frequency) and *Ceramic R&D Programs* (annual); and the Thermophysical Properties Research Center (D30), together with University Microfilms (E20), publishes *Masters Theses in the Pure and Applied Sciences Accepted by Colleges and Universities in the United States* (annual).

The National Science Foundation is one of the major sources of government research funds, and the *NSF Yearbook*[1] is a complete guide to NSF programs and activities. It includes information on grants and awards, legislation, programs, statistics, and personnel and is fully indexed. Another source for information on government research funds is the *Annual Register of Grant Support.*[2]

Universities: Although industrial companies perform much basic research in their laboratories, proprietary interests prohibit open publications describing their activities. Information on research performed in universities, on the other hand, is generally unclassified (except for classified defense projects), and guides are available. ASM's *Metallurgy/Materials Education Yearbook* (B15) gives the faculties and fields of interest of the 90 universities listed on pages 3–5. The American Chemical Society (B5) has a *Directory of Graduate Research*—a substantial volume, unfortunately without a subject index.

Another directory of university research is Wohlbier's *Worldwide Directory of Mineral Industries Education and Research* (G20), which has a section on "Metallurgy, Ceramics and Fuel Technology" and an excellent subject index. Gale Research Co. (E7) publishes a *Research Centers Directory* which includes other nonprofit organizations as well as universities. The subject index to the 1968 *Directory* (it is updated by periodical supplements) lists 52 such entries under the heading "Metallurgy."

University Microfilms (E20), which specializes in dissertations as well as microfilm editions of other publications, issues *American Doctoral Dissertations* annually; also *Masters Theses in the*

---

[1] *NSF Yearbook.* Academic Media, Inc., 32 Lincoln Ave., Orange, N.J. 07050. Annual.

[2] *Annual Register of Grant Support.* Academic Media, Inc., 32 Lincoln Ave., Orange, N.J. 07050. Annual.

Pure and Applied Sciences. Its alerting service, DATRIX, compiles bibliographies of dissertations in response to keywords provided by subscribers.

Miscellaneous: The Institute for Scientific Information (E10) issues an annual directory of the authors included in its broad coverage which is known as *ISI's Who is Publishing in Science* (no subject index); the American Institute of Consulting Engineers has a directory of engineering consultants;[3] and the American Chemical Society (B5) publishes a monthly journal, *Accounts of Chemical Research.*

A few of the association journals include sections on ongoing research. Three examples are the *Journal of the Electrochemical Society* (B33), the *Cast Metals Research Journal* (B8), and the *American Ceramic Society Bulletin* (B4).

Scientific meetings are extremely valuable sources of up-to-the-minute information. It would be well to watch your junk mail for notices that might easily be thrown away. The best formal alerting service for metallurgists is the *World Calendar of Forthcoming Meetings; Metallurgical and Related Fields* issued by the British Iron and Steel Institute (B46). More general coverage is provided by *Scientific Meetings,*[4] published by the Special Libraries Association, and by CCM's *World Meetings, U.S. and Canada* and *World Meetings, Outside U.S. and Canada* (E3). CCM also offers *Calls for Papers,* published weekly. A related service covering past meetings is *Proceedings in Print.*[5]

Supplementing CCM's service is a new publication called *Current Programs,*[6] issued monthly. It is divided into 17 subject sections, including 1 on "Materials Science and Technology." A magnetic tape service covering the *Current Programs* data files also is available.

---

[3] *Engineering Consultants Directory.* 5th ed. 1970. American Institute of Consulting Engineers, 345 E. 47th St., New York, N.Y. 10017.

[4] *Scientific Meetings.* Special Libraries Association, 235 Park Ave., S., New York, N.Y. 10003. Quarterly.

[5] *Proceedings in Print.* Edited by B. A. Spence. Proceedings in Print, Inc., P.O. Box 247, Mattapan, Mass., 02126. Bimonthly. Subscription $45.

[6] *Current Programs.* World Meetings Information Center, Inc., 824 Boylston St., Chestnut Hill, Mass. 02167. Monthly.

DIRECTORY

A

LIBRARIES

A1

AMES LABORATORY LIBRARY
U.S. Atomic Energy Commission
Iowa State University
Ames, Iowa 50012

HOLDINGS: 2,000 books; 180,000 government documents. Metallurgical interests are high-purity metals and rare-earth metals.

SERVICES: Open to the public for reference; limited interlibrary loan.

A2

ARGONNE NATIONAL LABORATORY
LIBRARY
9700 S. Cass Ave.
Argonne, Ill. 60439

HOLDINGS: 100,000 volumes; 1,800 journals and serials; 200,000 unclassified AEC reports. Principal holdings are in physical, theoretical, and nuclear metallurgy.

SERVICES: Open to the public for on-site use only; interlibrary loan.

A3

ATLANTIC REGIONAL LABORATORY
LIBRARY, NATIONAL RESEARCH COUNCIL OF CANADA
1411 Oxford St.
Halifax, Nova Scotia, Canada

HOLDINGS: Approximately 12,000 volumes. Principal metallurgical coverage is in extractive and chemical metallurgy.

SERVICES: Open to the public for reference; direct loan and interlibrary loan; photocopy service available on periodical articles and book chapters.

A4

BATTELLE-NORTHWEST TECHNICAL
LIBRARY
Richland, Wash. 99352
    See also BATTELLE MEMORIAL
    INSTITUTE (F1)

HOLDINGS: Approximately 40,000 books; 2,000 serials; 250,000 reports. Metallurgical interests include physical and mechanical metallurgy, irradiation effects on metals, nuclear metallurgy, metallography, and a limited amount of material on fabrication and forming.

SERVICES: Open to the public for official use; interlibrary loan and photocopy services; literature searches at an hourly fee by letter agreement.

A5

CARNEGIE LIBRARY OF PITTSBURGH
Science and Technology Dept.
4400 Forbes Ave.
Pittsburgh, Pa. 15213

HOLDINGS: 396,000 reference volumes, 85,000 circulating volumes, and 5,000 current periodicals in the collection of the Science and Technology Department. A free public library maintained by the city, county, and state. Service to industry is one of its prime goals. The science and technology of metals, with special emphasis on iron and steel, are important features of the collection.

SERVICES: Open to the public for reference and direct loan; interlibrary loan and photocopy services.

PUBLICATION: For more than 50 years, the Library has published the annual *Review of Iron and Steel Literature*, a classified list of the more important books, reports, and pamphlets issued during the year. It is highly selective and does not cover the periodical literature.

SPECIAL COLLECTIONS: A file of translations including those issued by Henry Brutcher, the British Iron and Steel Institute, U.S. Atomic Energy Commission, Consultants Bureau, and National Science Foundation; extensive collection of U.S. and British patents.

A6

ENGINEERING SOCIETIES LIBRARY
345 E. 47th St.
New York, N.Y. 10017

HOLDINGS: 225,000 volumes, plus more than 5,000 periodicals and other serial publications currently being received from about 50 countries in 25 languages. It is considered "the largest engineering library in the free world" and undoubtedly contains one of the most extensive metallurgical collections. Its holdings primarily are intended to serve the graduate and practicing engineer.

SERVICES: Open to the public for reference and photocopy service. Direct loan is available to members of the Library's 11 supporting technical societies, 3 of which have metallurgical interests: the American Institute of Mining, Metallurgical, and Petroleum Engineers; the American Society of Mechanical Engineers; and the American Welding Society.

Company memberships, available at a fee, include book-borrowing and other special privileges. Interlibrary loan is not provided.

PUBLICATIONS: Bibliographies are published irregularly and priced separately. A list is available on request.

SPECIAL SERVICES AND COLLECTIONS: Literature searches, ranging from a list of a few references for a specific request to the preparation of comprehensive annotated bibliographies, are offered at an hourly rate. A translation service is offered at standard rates per hundred words, covering all languages in which engineering documents are published. The Library is also the source and depository of all publications abstracted in *Engineering Index* (F4).

SELF-HELPS: *A Bibliography on Filing, Classification and Indexing Systems, and Thesauri for Engineering Offices and Libraries*, ESL Bibliography No. 15, 1966, 37 pp. Arranged in four parts: general information, universal systems, systems for special subjects, and thesauri; plus a subject index.

A7

JOHN CRERAR LIBRARY
35 W. 33rd. St.
Chicago, Ill. 60616

HOLDINGS: More than 1 million volumes and 12,000 periodicals and serials. A free public reference library devoted to research literature in the fields of science, technology, and medicine. Holdings are particularly strong in the basic sciences and in technology, including all branches of engineering and applications of the physical sciences in industry; thus, the metallurgical collection is one of the best.

SERVICES: Open to the public for reference; interlibrary loan limited to "members" (contributors to financial support); photocopy service.

PUBLICATIONS: *Translations Register-Index* (semimonthly). Announces and indexes all translations currently collected by the National Translations Center (see SPECIAL SERVICES AND COLLECTIONS below). This title, established in 1967, was preceded by *Technical Translations* (published by the Clearinghouse for Federal Scientific and Technical Information, 1959–1967) and *Translations Monthly* (sponsored by the Special Libraries Association at Crerar, 1955–1958). The translations are classified according to COSATI subject categories (see National Technical Information Service, C10). It includes a journal citation index, a list of translated patents arranged by country of origin, and lists of translated conference papers and monographs. Index sections also include translations available from other sources and are cumulated quarterly for all entries to date in a volume.

Consolidated Index of Translations into English. A compilation of most of the translations announced prior to the advent of *Translations Register-Index* in 1967. It can be purchased from the Special Libraries Association, 235 Park Ave., S., New York, N.Y. 10003.

Guide to Metals Literature. One of several guides to the Library's collections. It lists titles of reference works (including handbooks, dictionaries, statistical sources, directories, standards, and specifications), bibliographic sources (including indexing and abstracting services, bibliographies, indexes in periodicals, and literature guides), serials (including reports of conferences, society publications, periodicals, and reviews), and historical materials.

SPECIAL SERVICES AND COLLECTIONS: Retrospective literature searches are provided at an hourly rate; current awareness services are available at a negotiated monthly rate. Crerar houses and services the U.S. National Translations Center, a depository of more than 160,000 unpublished translations, and also acts as a clearinghouse on availability of other translations.

A8

LIBRARY OF CONGRESS
Science and Technology Division
Washington, D.C. 20540

HOLDINGS: More than 2 million books, more than 20,000 journals and other serial titles, and the most widely representative collection of technical reports in the country—approximately 1.25 million. (The total collections of the entire Library of Congress number 61 million items.)

SERVICES: Open to the public for reference only. Interlibrary loan requests should be addressed to the Library of Congress Loan Division, not to the Science and Technology Division; photocopy requests should be addressed to the Library's Photoduplication Service. Rates and conditions for both of these services are available on request.

PUBLICATIONS: Primarily bibliographies, checklists, journal lists, and similar reference tools. Only a few are of peripheral metallurgical interest, and some are quite old.

SPECIAL INFORMATION SERVICES: Answers reference questions, performs brief literature searches, compiles subject bibliographies on request, and offers advisory services; performs comprehensive literature searches at an hourly fee, in conjunction with the National Technical Information Service; operates the National Referral Center for Science and Technology (see C8).

A9

LINDA HALL LIBRARY
5109 Cherry St.
Kansas City, Mo. 64110

HOLDINGS: A science and technology reference library with some 500,000 items, including 25,000 journal and serial titles, half of which represent current subscriptions. More than 35 languages—predominantly English, Russian, Japanese, and German—are represented. The Library was "planned especially to serve the library needs of Kansas City and its territory . . ." and, since this area represents a diversified industrial complex, its metallurgical collections are broad and quite extensive. In 1970, the Library answered some 40,000 requests from the 50 states and 26 foreign countries for interlibrary loan, photocopies, and microfilm.

SERVICES: Open to the public for reference; interlibrary loan and photocopy services.

PUBLICATIONS: *The Book Collection and Services of the Linda Hall Library—an Outline Guide*, 1967, 48 pp. Contains general information on how to use the library, including patent searching and use of the card catalog.

A number of bibliographies and bulletins, in a variety of categories covering the Library's holdings, are also issued. Unfortunately, Bibliography No. 11, "Mining and Metallurgy," is out of date and out of print.

SPECIAL SERVICES: The Missouri State Technological Center, supported by the University of Missouri, is located in Linda Hall Library and uses its facilities for services to Missouri industry and business organizations.

A10

LOS ALAMOS SCIENTIFIC LABORATORY LIBRARIES
P.O. Box 1663
Los Alamos, New Mex. 87544

HOLDINGS: 110,000 books; 3,800 journals; 72,000 bound journals; 160,000 reports. Metallurgical interests are uranium, plutonium, and various "exotic" metals.

SERVICES: Open to the public for reference; interlibrary loan and limited photocopy service.

A11

MINES LIBRARY, DEPARTMENT OF NATURAL RESOURCES (MINISTÈRE DES RICHESSES NATURELLES)
Parliament Bldg.
1640 Blvd. de l'Entente
Quebec , P.Q., Canada

HOLDINGS: 50,000 volumes; 450 subscriptions. All aspects of metallurgy are covered.

SERVICES: Open to the public for reference; interlibrary loan and photocopy services; occasional literature searches on request. Catalog of the library, with index, is in preparation.

A12

NATIONAL SCIENCE LIBRARY OF CANADA
National Research Council of Canada
Ottawa 7, Ont., Canada

HOLDINGS: More than 800,000 books, bound periodicals, pamphlets, and technical reports; approximately 16,500 scientific and technical journals published throughout the world; complete files of scientific and technical abstracting and indexing services; depository collections of U.S. Government reports; a comprehensive collection of unilingual, bilingual, and multilingual dictionaries covering all fields of science and technology. All aspects of metallurgy are covered.

SERVICES: Open to the public for reference; interlibrary loan and photocopy services.

PUBLICATIONS: NRC's *Technical Translation* series, distributed on an exchange basis to scientific organizations, libraries, and to any interested Canadian scientists.

INFORMATION RETRIEVAL: CAN/SDI Program is a national computer-based current awareness—Selective Dissemination of Information (SDI) —service. Resources include Chemical Abstracts Service's *CA* Condensates and *Chemical Titles* (F3), Institute for Scientific Information's Citation File tapes (E10), and INSPEC tapes (B40).

The Library also operates a "Question and Answer Service" designed to supplement local and regional facilities at no charge. Requests should be submitted through a local library or information center; requests from outside Canada normally are limited to those concerning Canadian material.

SPECIAL SERVICES: NSL's Translations Section maintains the *Canadian Index of Scientific Translations*, which records the location of more than 200,000 translations of foreign scientific papers prepared by commercial agencies and private translators in all parts of the world. The Library provides, or assists in obtaining, copies of desired translations.

A13

ONTARIO DEPARTMENT OF MINES LIBRARY
Parliament Bldg.
Queen's Park, Zone 2
Toronto, Ont., Canada

HOLDINGS: Approximately 500 books and 125 serials; all publications of the Department. Metallurgical coverage is minor, with emphasis on geology and mining.

SERVICES: Open to the public for reference; interlibrary loan and photocopy services.

A14

ONTARIO RESEARCH FOUNDATION LIBRARY
43 Queen's Park, Zone 5
Toronto, Ont., Canada

HOLDINGS: 9,000 books; 500 journals and serials. Subjects include chemistry, engineering, metallurgy, and physics.

SERVICES: Open to the public for reference; interlibrary loan; photocopy service on periodical articles and book chapters.

A15

SOUTHERN RESEARCH INSTITUTE LIBRARY
2000 Ninth Ave., S.
Birmingham, Ala. 35205

HOLDINGS: Approximately 29,000 bound volumes; 800 periodical subscriptions. Metallurgical coverage is primarily in the fields of casting, chemical properties, and analytical techniques.

SERVICES: Open to the public for reference; interlibrary loan and photocopy services.

A16

SOUTHWEST RESEARCH INSTITUTE LIBRARY
8500 Culebra Rd.
San Antonio, Tex. 78206

HOLDINGS: Approximately 25,000 volumes; more than 1,000 current periodicals and serials. Primary interests are in the inorganic science and engineering fields. All aspects of metallurgy are covered, but from a research rather than a production standpoint.

SERVICES: Open to the public for reference; interlibrary loan and photocopy services.

B

ASSOCIATIONS

B1

ALUMINIUM FEDERATION (AF)
Hastings House
10 Norfolk St.
London, W.C.2, England

PURPOSE AND FUNCTION: A trade association for the British aluminum industry.

PUBLICATIONS: Some 25 information bulletins (separate pamphlets dealing with various technical aspects of aluminum fabrication and use), plus research reports, application brochures, and symposium proceedings. A list of publications is available on request.

Also serves as European editorial office of *World Aluminum Abstracts,* representing the Centre International de Développement de l'Aluminium (B29).

B2

THE ALUMINUM ASSOCIATION (AA)
750 Third Ave.
New York, N.Y. 10017

PURPOSE AND FUNCTION: A trade association representing the primary producers of aluminum in the United States, leading manufacturers of semifabricated products, and principal foundries and smelters.

PUBLICATIONS:
Abstracts and Indexes: World Aluminum Abstracts (monthly,

March 1968–). Sponsored jointly with Centre International de Développement de l'Aluminium and prepared by the American Society for Metals. For details, see B15.

Index to Publications Related to Aluminum Extrusions. Published by the Aluminum Extruders Council. A small brochure listing periodical articles, industrial pamphlets, and publications of The Aluminum Association.

Standards: Aluminum Standards and Data.

Also provides a number of miscellaneous standards dealing with various aspects of design and finishing.

Other: Nearly 40 manuals, primarily application- and design-oriented.

SELF-HELPS: *Thesaurus of Aluminum Technology* (in press). Provides a standardized indexing vocabulary and is used as the authority for the index in *World Aluminum Abstracts.*

B3

ALUMINUM SMELTERS RESEARCH INSTITUTE (ASRI)
20 N. Wacker Dr.
Chicago, Ill. 60606

PURPOSE AND FUNCTION: A trade association for companies producing secondary aluminum. ASRI has no formal library, but maintains a card index at headquarters covering publications on secondary aluminum and its applications, with concentration on die casting.

B4

AMERICAN CERAMIC SOCIETY (ACS)
4055 N. High St.
Columbus, Ohio 43214

PURPOSE AND FUNCTION: A professional society of scientists, engineers, plant operators, and others interested in glass, ceramic-metal systems, refractories, white wares, electronics, and structural clay products industries. Its purpose is to promote methods of research and production in the ceramics field.

LIBRARY HOLDINGS: Approximately 1,100 volumes.

LIBRARY SERVICES: Open to the public for reference; photocopy service.

PUBLICATIONS:
Primary Journals: American Ceramic Society Bulletin (monthly). Includes technical papers and news. Each February, publishes a list of ceramic research projects in progress.
Journal of the American Ceramic Society (monthly). Consists primarily of research papers, discussions, and notes.
Books: Engineering Properties of Selected Ceramic Materials, 675 pp. A materials selection databook prepared by Battelle Memorial Institute Columbus Laboratories under contract to the U.S. Air Force.
Refractory Ceramics for Aerospace. A revised, enlarged, and updated edition published in 1964.
1969 Supplement—Phase Diagrams for Ceramists, 625 pp. Adds 2,083 diagrams to the 2,066 contained in the original volume published in 1964.
A series of four *Special Publications*, mainly symposia and compilations, and two *Proceedings* volumes. Order list available on request.
Abstracts: Ceramic Abstracts (monthly, 1918–). Appears as a section in the *Journal of the American Ceramic Society*. Provides about 7,500 abstracts per year from 500 worldwide journals and other sources. Arranged in 17 subject sections with annual subject and author indexes.
Ceramic-Metal Systems and Enamel Bibliography and Abstracts (annual, 1928–). Provides about 425 abstracts per year from 125 journals, plus patents, books, and conference papers; includes author and subject indexes.

B5

AMERICAN CHEMICAL SOCIETY (ACS)
1155 16th St., N.W.
Washington, D.C. 20036
(see also CHEMICAL ABSTRACTS SERVICE, a division of ACS, F3)

PURPOSE AND FUNCTION: A scientific, educational, and professional society formed to provide leadership in advancing many aspects of chemical research, technology, and education. Of the 26 technical divisions, 4 have a certain amount of metallurgical interest: Division of Analytical Chemistry, Division of Inorganic Chemistry, Division of Physical Chemistry, and Division of Industrial and Engineering Chemistry.

LIBRARY HOLDINGS: 6,000 volumes and 500 periodical titles. All ACS journals are available on microfilm, including current and back issues.

LIBRARY SERVICES: Open to the public for reference; interlibrary loan; photocopy service confined to ACS journals.

PUBLICATIONS:
Primary Journals: Of the 19 ACS journals, the following carry occasional articles and papers of metallurgical interest:
Analytical Chemistry (monthly).
Chemical & Engineering News (weekly).
Chemical Technology (monthly).
Industrial and Engineering Chemistry in three separate sections:
I&EC—Process Design and Development (quarterly).
I&EC—Product R&D (quarterly).

I&EC—Fundamentals (quarterly).
Journal of Chemical and Engineering Data (quarterly).
Journal of Physical Chemistry (biweekly).
Reviews and Indexes: Chemical Reviews (bimonthly). Provides critical and comprehensive reviews on various topics.
Accounts of Chemical Research (monthly). Provides critical reviews of current active investigations.
Journal of Physical and Chemical Reference Data (quarterly, 1972–). Published jointly with the American Institute of Physics. Compilations of evaluated reference data and critical reviews prepared under the program of the National Bureau of Standards' National Standard Reference Data System (C7).
Annual Index to Chemical & Engineering News.
Other: Directory of Graduate Research. Lists research programs being carried out by faculty members of colleges and universities in the United States and Canada. Arrangement is alphabetical by name of institution. Published in four sections: Chemistry, Chemical Engineering, Biochemistry, and Pharmaceuticals. Since there is no subject index, it would be difficult to search for those engaged in metallurgical research.

SPECIAL SERVICES: Single Article Announcement Service. A biweekly current awareness service that reproduces contents pages from current issues of most of the ACS primary journals (excludes *Chemical Reviews* and *Accounts of Chemical Research*). Articles of special interest can be ordered without subscribing to the complete journal.

B6

AMERICAN DIE CASTING INSTITUTE (ADCI)
366 Madison Ave.
New York, N.Y. 10017

PURPOSE AND FUNCTION: A trade association for custom producers of die castings made of zinc, aluminum, magnesium, and brass.

PUBLICATIONS: *Production Standards for Die Castings,* 50 pp. Arranged in three series: "Engineering," "Metallurgical," and "Commercial."

B7

AMERICAN ELECTROPLATERS' SOCIETY (AES)
56 Melmore Gardens
East Orange, N.J. 07017

PURPOSE AND FUNCTION: A professional society of scientists, technicians, and others interested in research on electroplating, deposition and finishing of metals, and allied arts.

LIBRARY SERVICES: A small library is open to the public for reference.

PUBLICATIONS: *Plating* (monthly). Patents abstracts section appears in each issue.
Occasional books of symposia and conference proceedings.

B8

AMERICAN FOUNDRYMEN'S SOCIETY (AFS)
Golf and Wolf Rds.
Des Plaines, Ill. 60016

PURPOSE AND FUNCTION: A technical society of engineers, metal-casting scientists, technologists, foundrymen, pattern-makers, and educators. Divisions include: Brass and Bronze, Casting Design, Ductile Iron, Education, Gray Iron, Light Metals, Malleable Iron, Molding Methods and Materials, Pattern-making, Plant Engineering, Sand, Steel, and Environmental Control.

LIBRARY HOLDINGS: Approximately 3,500 volumes, plus most journals dealing with various aspects of metal-casting and foundry technology; conference and meeting papers; translations from some French, German, and Russian journals.

LIBRARY SERVICES: Open to the public for reference; interlibrary loan and photocopy services.

PUBLICATIONS:
Primary Journals: Modern Casting (monthly).
Cast Metals Research Journal (quarterly).

Transactions of the American Foundrymen's Society (annual).

Books: Cast Metals Handbook; Foundry Sand Handbook; plus nearly 100 textbooks, symposia, research reports, and training manuals.

INFORMATION RETRIEVAL SERVICES: Three services are offered:

1. Document Retrieval Service. Designed to provide, on demand, technical information selected from an extensive collection of references, dating back to 1950, which are cataloged on retrieval cards showing title, author, and citation. When a specific subject request is received, a bibliography is compiled and mailed to the requester, who can then order original copies of desired documents from the AFS Technical Information Center.

2. Current Awareness Service. Provides reference cards containing informative abstracts of materials available in the library. Each card carries a subject heading at the top for filing purposes. Cards are mailed twice a month, with about 40 cards per mailing. The Service includes a card-filing classification system, a file box, index tab cards, and an annual catalog.

3. Special Literature Search Service (an adjunct to the Document Retrieval Service). Available at an hourly rate, when the text of numerous articles and books must be examined in order to answer a request.

B9

AMERICAN HOT DIP GALVANIZERS ASSOCIATION (AHDGA)
1000 Vermont Ave., N.W.
Washington, D.C. 20005

PURPOSE AND FUNCTION: A trade association designed to promote the hot-dip galvanizing industry through advertising and public relations, technical services and development, advice on management problems, and market development. Metallurgical interests cover steel coating and corrosion prevention.

PUBLICATIONS: Approximately 40 reports and promotional brochures; recommended procedures.

SPECIAL SERVICES: Consultation services are primarily for members, but also are available to nonmembers as time and information permit.

B10

AMERICAN INSTITUTE OF AERONAUTICS AND ASTRONAUTICS (AIAA)
Technical Information Service
750 Third Ave.
New York, N.Y. 10017

PURPOSE AND FUNCTION: A society for the interchange of technological information through publications and technical meetings. AIAA operates primarily through 38 committees. Metallurgical interests include physical and mechanical properties of materials, structural applications of materials, corrosion, and cermets.

LIBRARY HOLDINGS: 10,000 books; 40,000 reports and papers; 100,000 microfiche.

LIBRARY SERVICES: Open to the public for reference; photocopy service; literature searches on demand.

PUBLICATIONS:

Primary Journals: Six titles, two with occasional articles of metallurgical interest: *AIAA Journal* (monthly) and *Astronautics and Aeronautics* (monthly).

Abstracts: International Aerospace Abstracts (semimonthly, 1961 –). About 19,000 abstracts per year from about 850 worldwide journals. Subject, author, report number, and accession number indexes appear in each issue and are cumulated quarterly and annually. This is a companion publication to *Scientific and Technical Aerospace Reports* (NASA, C5) issued on alternate weeks, with joint cumulative indexes.

Other: Specialized technical reports, technical meeting papers, and proceedings.

INFORMATION RETRIEVAL SERVICES: Magnetic tapes of *International Aerospace Abstracts* and *Scientific and Technical Aerospace Reports* are prepared by NASA and used as the basis for its information retrieval services (see C5). The tapes also are made available to the NASA Regional Dissemination Centers.

B11

AMERICAN INSTITUTE OF CHEMICAL ENGINEERS (AIChE)
345 E. 47th St.
New York, N.Y. 10017

PURPOSE AND FUNCTION: A professional society of chemical engineers, AIChE has a number of committees and divisions, including a division on Materials Engineering and Sciences.

LIBRARY: The facilities of the Engineering Societies Library (A6), housed in the same building, are used.

PUBLICATIONS:
Chemical Engineering Progress (monthly).
AIChE Journal (bimonthly).
International Chemical Engineering (quarterly).
Monograph and symposia series volumes, technical manuals, reprint manuals, and standards.

SELF-HELPS: Abstracts, plus keyword index terms for each paper or article in the Institute's journals, are printed so that they can be clipped and mounted on cards. Keywords are based on the *Chemical Engineering Thesaurus* published by AIChE in 1961. Information Retrieval Cards can also be purchased by an individual to organize his files according to the AIChE concept coordination system of information storage and retrieval.

B12

AMERICAN INSTITUTE OF PHYSICS (AIP)
335 E. 45th St.
New York, N.Y. 10017

PURPOSE AND FUNCTION: A federation of leading societies in the field of physics, AIP combines into one operating agency those functions that can best be accomplished jointly. Its purpose is advancing and disseminating the knowledge of physics and its application for human wel-

fare. There are 7 member societies and 19 affiliated societies of which 2—the American Physical Society (which has nine divisions including one on solid-state physics) and the American Crystallographic Association—are of metallurgical interest. The American Vacuum Society (B19) is an affiliate member.

LIBRARY HOLDINGS: Approximately 8,000 volumes, including back issues of all AIP journals, in the Niels Bohr Library for the History and Philosophy of Physics, maintained by AIP.

LIBRARY SERVICES: Open to the public for reference; direct and interlibrary loan; limited photocopy service.

PUBLICATIONS:
Primary Journals: Eighteen archival journals; 5 society bulletins and programs (a number of which are sponsored by its various member societies); a monthly news magazine, *Physics Today*; and 13 cover-to-cover translations of Russian journals. Those containing a substantial quantity of metallurgical information are:

Physical Review B (Solid State)
Physical Review Letters
Applied Spectroscopy
Journal of Vacuum Science and Technology
Review of Scientific Instruments
Journal of Applied Physics
Applied Physics Letters
Soviet Physics—Crystallography
Soviet Physics—JETP
JETP Letters
Soviet Physics—Solid State
Soviet Physics—Semiconductors
Soviet Physics—Technical Physics.

One additional translated Soviet journal, *High Temperature*, is published in cooperation with the Consultants Bureau. These journals, including back issues, are also available on microfilm.

AIP distributes the publications of the Institute of Physics and the Physical Society (London) in North and South America. Journals of interest to metallurgists include the *Journal of Physics E* (Scientific Instruments) and *Journal of Physics F* (Metal Physics).

Books: The *AIP Handbook* (published by McGraw-Hill). Contains 2,000 pages of physical data.

Temperature; Its Measurement and Control (published by Reinhold). Three-volume reference book.

Occasional conference proceedings are also of metallurgical interest.

Abstracts, Indexes, and Reviews:

Current Physics Advance Abstracts (CPAA) (monthly, 1972–). Contains abstracts of papers of all AIP-published archival journals and translations, distributed as early as two months before publication of the complete papers. In three sections, sold separately: *CPAA Solid State, CPAA Nuclei & Particles,* and *CPAA Atoms & Waves.*

Current Physics Titles (CPT) (monthly, 1972–). Provides subject index to more than 60 journals covering approximately 2,000 articles and papers each month. In three sections, sold separately: *CPT Solid State, CPT Nuclei & Particles,* and *CPT Atoms & Waves.* Arranged by subject, with each article citation appearing as many times as it has classification numbers. *CPT* is essentially a printed version of SPIN, AIP's computer-readable magnetic tape service described below.

AIP is also the U.S. agent for servicing subscriptions to *Physics Abstracts* and *Current Papers in Physics,* published by the Institution of Electrical Engineers (B40).

Another joint enterprise is the *Journal of Physical and Chemical Reference Data,* a critical review publication for which the American Chemical Society (B5) acts as subscription agent.

SPECIAL SERVICES: Searchable Physics Information Notices (SPIN) (annual, updated monthly). A magnetic tape service with the same coverage as *Current Physics Titles.* Tapes contain article title, authors and their affiliations, abstract, journal citation, special index terms and keywords, and references to all cited journal articles. Tapes may be leased for in-house searches. Retrieval services also are available from licensed information centers.

B13

AMERICAN METAL STAMPING ASSOCIATION (AMSA)
3673 Lee Rd.
Cleveland, Ohio 44120

PURPOSE AND FUNCTION: A trade association of metal stamping manufacturers and their suppliers. It conducts research and educational activities in cooperation with colleges and universities.

LIBRARY HOLDINGS: Approximately 250 books, 500 catalogs and pamphlets, and 100 journals and serials.

PUBLICATIONS:
Metal Stamping (bimonthly).
Stamping Guide (annual).
Standards and occasional books.

B14

AMERICAN SOCIETY OF MECHANICAL ENGINEERS (ASME)
345 E. 47th St.
New York, N.Y. 10017

PURPOSE AND FUNCTION: A professional society of mechanical engineers, ASME conducts research; develops boiler, pressure vessel, and power test codes; and develops equipment safety codes and standards. Metallurgical coverage is marginal.

LIBRARY SERVICES: As one of the supporting members, uses the facilities of the Engineering Societies Library (A6), housed in the same building.

PUBLICATIONS:
Primary Journals: Mechanical Engineering (monthly).
Quarterly Transactions of the ASME. Published as seven separate journals, four of which have some metallurgical interest: *Journal of Applied Mechanics, Journal of Basic Engineering, Journal of Engineering for Industry,* and *Journal of Lubrication Technology.*
Books: The current catalog lists titles in 14 subject categories; 20 titles are listed under "Materials/Metals," including 4 *ASME Handbooks: Engineering Tables, Metals*

Engineering—Design, Metals Engineering—Processes, and *Metals Engineering—Metals Properties.* All were published prior to 1960 except *Design,* published in 1965.

Abstracts and Reviews: Applied Mechanics Reviews (monthly). Prepared at Southwest Research Institute. More than 10,000 critical abstracts annually, taken from world literature. Arranged in 5 main sections and 54 subsections, the latter including "Material Processing," "Fracture (Including Fatigue)," "Experimental Stress Analysis," and "Materials Test Techniques." Also includes extended critical reviews, with accompanying bibliographies, and annual subject and author indexes.

Standards and Specifications: A large number of standards and codes, most of which are primarily of mechanical interest. The *ASME Boiler and Pressure Vessel Code* includes two sections of *Materials Specifications: Part A—Ferrous* and *Part B—Nonferrous.*

B15

AMERICAN SOCIETY FOR METALS
(ASM)
Attn: Metals Information
Metals Park, Ohio 44073

PURPOSE AND FUNCTION: A nonprofit educational and technical society dedicated to advancing scientific, engineering, and technical knowledge, particularly with respect to the manufacture, treatment, selection, and use of metals and other engineering materials. Goals are accomplished through education, research, and the compilation and dissemination of information. The subject area encompasses the manufacture, treatment, selection, and use of metals and related engineering materials. Metals coverage is further indicated by the following titles of present and proposed technical divisions: materials science, materials systems and design, metals production, heat treatment, mechanical working and forming, machining and tooling, welding and joining, cleaning and finishing, casting processes and foundry technology, materials testing and quality control, and materials in processing equipment.

LIBRARY HOLDINGS: 6,000 books; 1,000 serials; microfilm of journal articles dating back to 1959; and miscellaneous pamphlets, reports, reprints, and preprints.

LIBRARY SERVICES: Not open to the public. Interlibrary loan; photocopy service, based on microfilm of most of the journal articles abstracted in *Metals Abstracts* and its predecessor, *Review of Metal Literature,* dating back to 1959—approximately 150,000 items. The library also maintains the Eisenman Rare Book Collection containing 43 items.

PUBLICATIONS:
Primary Journals:
Metal Progress (monthly).
ASM News Monthly.
Metallurgical Transactions (monthly); published jointly with The Metallurgical Society of AIME.
Metals Engineering Quarterly.
Materials Science and Engineering (monthly); published jointly with Elsevier Sequoia, Switzerland.
Books:
Metals Handbook, 8th ed., in 10 volumes. Published to date:

Vol. 1—Properties and Selection of Metals, 1961, 1,300 pp.
Vol. 2—Heat Treating, Cleaning and Finishing, 1964, 708 pp.
Vol. 3—Machining, 1967, 552 pp.
Vol. 4—Forming, 1969, 528 pp.
Vol. 5—Forging and Casting, 1970, 472 pp.
Vol. 6—Welding, 1971, 734 pp.
Vol. 7—Atlas of Microstructures of Industrial Alloys, 1972, 366 pp.

Forthcoming volumes:

Vol. 8—Failure Analysis.
Vol. 9—Nondestructive Testing and Inspection.
Vol. 10—Metals Engineering Design.

Casting Design Handbook, 1962, 326 pp. Sponsored jointly by ASM and the U.S. Air Force.

Forging Design Handbook, 1972, 308 pp. Sponsored by the U.S. Air Force.

Metal Progress Databook, 2nd ed., 1970, 164 pp., looseleaf. A collection of data sheets for the areas of material selection, process engineering, and fabrication technology.

For the Atomic Energy Commission, Division of Technical Information, ASM has directed the preparation of 13 books in a *Monograph Series* on "Metallurgy in Nuclear Technology." These are listed in the AEC catalog of *Technical Books and Monographs* (C15).

Approximately 45 other miscellaneous ASM technical books are now in print. A catalog/price list is available on request.

Abstracts and Indexes:

Metals Abstracts (monthly, 1968–). Published jointly with the Institute of Metals, London. About 25,000 informative abstracts per year —in 33 subject categories—taken from about 1,000 worldwide journals, plus books, reports, and conference proceedings.

Metals Abstracts Index (monthly). Companion publication of *Metals Abstracts*. Provides subject and author indexes to the abstracts. Subscription includes annual cumulation as the 13th issue.

Metals Abstracts Annual Bound Volume. Monthly issues of *Metals Abstracts* are reformatted sequentially for each subject category. Includes annual index cumulation and list of journals abstracted.

Review of Metal Literature. Annual volumes, 1944–1967 inclusive. Predecessor publication to *Metals Abstracts.* Volumes vary in size and price.

Metalforming Digest (monthly, June 1972–). Approximately 1,800 abstracts per year on metal fabrication, including powder metallurgy.

World Aluminum Abstracts (monthly, March 1968–). An international journal prepared by ASM for The Aluminum Association, New York, and Centre International de Développement de l'Aluminium,

Paris. Known as *Aluminum Technical Information Service* through March 1970. About 5,500 abstracts per year in 38 subject categories. Coverage includes government reports and patents in addition to journals, books, and conference proceedings. Each issue includes subject and author indexes, and subscription includes annual index cumulation as a 13th issue. Bound volumes also are available.

List of Journals Abstracted by Metals Abstracts. Available on request.

Bibliographies: A series of 33 titles, sold individually, on practical metalworking topics. Each bibliography averages 300 to 400 references to the world literature. Titles are updated periodically. List available on request.

Reviews: International Metallurgical Reviews (quarterly, March 1972–). Published jointly with the Institute of Metals, London. Provides authoritative critical reviews of various aspects of metallurgical science and technology.

Other: ASM Report System Papers. Copies of individual papers presented at ASM conferences and meetings not published in the journals. The system currently offers 245 titles. Sets of papers presented at seminars and sessions on specific topics frequently are published in book form.

INFORMATION RETRIEVAL SERVICES: Both current awareness and retrospective literature searches. The data base is comprised of METADEX (the *Metals Abstracts Index* computer tapes) and *Review of Metal Literature* indexes dating back to the beginning of 1966, totaling about 125,000 items. The tape is updated monthly (average, 2,000 items per month) and contains complete citations to the original documents (average, 6 to 8 index terms) and serial numbers of corresponding abstracts. Tapes are also leased to other information centers and to industry. ASM searches are performed primarily by computer, but can be performed manually, depending on the nature of the question.

SELF-HELPS: The *ASM Thesaurus of Metallurgical Terms* is the vocabulary authority for the indexes to *Metals Abstracts* and *Review of Metal Literature*. It is useful for subscribers to the Information Retrieval Service and the METADEX tapes, as well as for personal indexing of other metallurgical material.

ASM-SLA Metallurgical Literature Classification. Out of print, but copies are available from University Microfilms (E20). The *Classification* can be used in filing documents or for card-indexing documents. Marginal hand-sorted punched cards, designed by ASM specifically for use with the *Classification*, are available from E-Z Sort Systems Parkison Agency, 35 E. Wacker Dr., Chicago, Ill. 60601.

Special indexes to non-ASM publications, as well as to special abstract collections, can be produced to order.

EDUCATIONAL AIDS: Metals Engineering Institute courses are offered on 22 metals subjects for home study; and in-plant and intensive training also are given. The Education Department also issues a *Metallurgy/Materials Education Yearbook* which lists schools with metallurgy facilities, together with names of professors and their research specialties.

B16

AMERICAN SOCIETY FOR NONDESTRUCTIVE TESTING (ASNT)
914 Chicago Ave.
Evanston, Ill. 60202

PURPOSE AND FUNCTION: A technical society for metallurgists, quality control specialists, welding engineers, industrial management personnel, technicians, and suppliers of equipment and services.

LIBRARY HOLDINGS: Approximately 800 volumes on radiography, magnetic testing, ultrasonics, and related subjects.

LIBRARY SERVICES: Open to the public for reference; interlibrary loan and photocopy services. All technical articles published in back issues of ASNT journals are available on microfilm.

PUBLICATIONS:
Primary Journal: Materials Evaluation (monthly).
Books: Nondestructive Testing Handbook. Edited and published for ASNT by Ronald Press.
Approximately 30 textbooks and conference proceedings; a number of programmed instructions and classroom training handbooks.
Abstracts: Current Literature on Nondestructive Testing. Provides abstracts published as a section of each issue of *Materials Evaluation*.
What's New in Nondestructive Testing Patents. A list of patents—most including summaries—also published as a section in *Materials Evaluation*.
Standards: Five books of recommended practices, each covering one common method of nondestructive testing.

B17

AMERICAN SOCIETY FOR QUALITY CONTROL (ASQC)
161 W. Wisconsin Ave.
Milwaukee, Wisc. 53203

PURPOSE AND FUNCTION: A nonprofit educational and scientific society concerned with quality control, reliability, inspection, research and development, statistics, and engineering. Four of its nine technical divisions reflect some interest in quality control as it is applied to metallurgical processing: Aircraft and Missile Division, Automotive Division, Chemical Division, and Inspection Division.

LIBRARY SERVICES: A small library is available for on-site reference only; photocopy service.

PUBLICATIONS:
Primary Journals:
Quality Progress (monthly).
Journal of Quality Technology (quarterly).
Technometrics (quarterly); pub-

lished jointly with the American Statistical Association.

Transactions of ASQC (annual).

Books: Processes orders for the *Quality Control Handbook* published by McGraw-Hill.

Indexes: A 25-year index to the *Transactions* and other ASQC periodical literature is in preparation.

Other: Three standards for definitions, symbols, and glossary; miscellaneous reports and reprints. List available on request.

B18

AMERICAN SOCIETY FOR TESTING AND MATERIALS (ASTM)
1916 Race St.
Philadelphia, Pa. 19103

PURPOSE AND FUNCTION: An international nonprofit, technical, scientific, and educational society devoted to promoting knowledge of the materials of engineering and the standardization of specifications and methods of testing. The Society operates through more than 105 technical committees, each having 5 to 10 subcommittees. Of the 33 volumes comprising the 1970 edition of the *ASTM Book of Standards*, 11 are of metallurgical interest.

LIBRARY HOLDINGS: Approximately 1,000 volumes on materials and standards, as well as selected journals.

LIBRARY SERVICES: Open to the public for reference only; photocopy service.

PUBLICATIONS:

Primary Journals:

Materials Research and Standards (monthly).

Journal of Materials (quarterly).

ASTM Proceedings (annual).

Books: The most renowned publication is the *ASTM Book of Standards* noted above. A new edition is issued annually.

Special reports, proceedings, and symposia are published in book form. The current catalog contains 67 titles of metallurgical interest.

Abstracts, Indexes, and Bibliog-

raphies: ASTM has maintained a *Bibliography on Fatigue* for many years, issued at two- to five-year intervals. Other bibliography topics are "Metal Cleaning" and "Thermostat Metals." Five-year indexes to ASTM technical reports are also issued.

SELF-HELPS: *Manual on Methods for Retrieving and Correlating Technical Data.*

B19

AMERICAN VACUUM SOCIETY (AVS)
335 E. 45th St.
New York, N.Y. 10017

PURPOSE AND FUNCTION: A technical society that seeks to advance and disseminate knowledge concerning vacuum science and engineering. One of its four divisions—Vacuum Metallurgy Division—is concerned with the processing and characteristics of metals under all subatmospheric conditions, including vacuum melting, casting, heat-treating, and purification.

PUBLICATIONS:

Primary Journal: Journal of Vacuum Science and Technology (bimonthly). Published by the American Institute of Physics under AVS's editorial management.

Other: A number of proposed vacuum standards, mostly in the field of vacuum technology.

B20

AMERICAN WELDING SOCIETY (AWS)
2501 N.W. 7th St.
Miami, Fla. 33125

PURPOSE AND FUNCTION: A professional engineering society devoted to welding and allied processes such as brazing, soldering, and hard facing.

LIBRARY SERVICES: Uses the facilities of the Engineering Societies Library (A6), housed in the same building.

PUBLICATIONS:

Primary Journal: Welding Journal (monthly).

Books: Welding Handbook in six volumes.

Welding Metallurgy, Vol. 1—Fundamentals; Vol. 2—Technology. About half a dozen other books.

Bibliographies and Indexes: AWS Bibliographies, 1937–1967; *Supplement,* 1968–1969; a list of articles published in the *Welding Journal,* classified in 1 or more of 37 subject categories.

Standards: Approximately 75 manuals of standards, specifications, and recommended practices.

B21

ASSOCIATION OF IRON AND STEEL ENGINEERS (AISE)
1010 Empire Bldg.
Pittsburgh, Pa. 15222

PURPOSE AND FUNCTION: A technical society designed to serve engineers and operators in the basic steel industry and suppliers to the industry. Its purpose is to advance the technical and engineering phases of iron and steel production and processing. Coverage emphasizes design and equipment aspects of the industry, but basic iron- and steel-producing metallurgy is well covered.

PUBLICATIONS: *Iron and Steel Engineer* (monthly).

Proceedings of the Association (annual).

Approximately half a dozen books; most are collections of papers or research reports.

AISE Standards (13 as of 1970) which primarily cover design and equipment specifications.

B22

BRITISH CAST IRON RESEARCH ASSOCIATION (BCIRA)
Alvechurch, Birmingham, England

PURPOSE AND FUNCTION: Performs research and provides advisory services to members and to users of cast iron. Maintains a library that claims to house the world's largest specialized collection on the ironfounding industry.

PUBLICATIONS: Primary publications describing Association work are available only to members.

Russian Castings Production, available on subscription basis (monthly). A cover-to-cover translation of the Russian journal, *Liteinoe Proizvodstvo.*

BCIRA Abstracts of Foundry Literature (bimonthly). Provides approximately 1,600 abstracts per year from 350 journals, as well as books, reports, proceedings, patents, and other material. Arranged in 12 main groups with 27 subgroups. Author index in each issue; subject and cumulated author indexes annually.

Compiles bibliographies; issues pamphlets of engineering data on various types of cast irons; in 1967, published *Chemical Analysis for Iron Foundries; Selected Methods.*

B23

BRITISH CERAMIC SOCIETY (BCS)
Shelton House
Stoke-on-Trent, England

PURPOSE AND FUNCTION: A technical society for the ceramics industry.

PUBLICATIONS: *Journal of the British Ceramic Society* (monthly).

Transactions of the British Ceramic Society (monthly). Contains *British Ceramics Abstracts* as a section.

Approximately 4,000 to 4,500 abstracts from journals, books, proceedings and conference volumes, reports, and patents are published each year in 19 subject categories. Abstracts are prepared by the British Ceramic Research Association.

A number of *Proceedings* volumes on various topics are also available.

B24

BRITISH IRON AND STEEL RESEARCH ASSOCIATION (BISRA)
24, Buckingham Gate
London, S.W.1, England

PURPOSE AND FUNCTION: Serves as the Inter-Group Laboratories of the British Steel Corp. In addition to facilities at the London headquarters, laboratories are operated in four other major steelmaking areas in the United Kingdom. BISRA, concerned with every aspect of iron and steel metallurgy from ore to delivery of steel in finished or semi-finished form, has five divisions: Iron Making, Steel Making, Mechanical Working, Plant Engineering and Energy, and Metallurgy.

PUBLICATIONS: Those restricted to members include: *Research Reports* (restricted when first issued; many are later reprinted for general distribution); *Members' Report List* (bimonthly); *Project Index* (annual); and *Closed Conference Proceedings.*

Reports available to the public are listed in the *Open Report List* (semiannual), available free on request. *BISRA Summaries* of selected research and *Open Conference Proceedings* also are generally available.

A complete list of miscellaneous publications, including bibliographies, is available on request.

B25

BRITISH NON-FERROUS METALS RESEARCH ASSOCIATION (BNF)
Euston St.
London, N.W.1, England

PURPOSE AND FUNCTION: Maintains laboratories for members' research and sponsors some work outside the Association; assists members in technical problems; and provides consultation service. BNF claims to "cover the world's metallurgical literature and provide any required data available in the literature or other sources."

PUBLICATION: *Bulletin* of the Association (available to members only).

B26

CANADIAN COPPER & BRASS DEVELOPMENT ASSOCIATION (CCBDA)
55 York St.
Toronto 117, Ont., Canada

PURPOSE AND FUNCTION: A nontrading, nonprofit organization sponsored by the Canadian copper industry to promote and develop the use of copper, its alloys, and compounds.

LIBRARY HOLDINGS: Approximately 500 books; 65 periodicals; and special collections of papers and research reports.

LIBRARY SERVICES: Limited to Canadian inquirers.

PUBLICATIONS: *Canadian Copper* (quarterly).

Approximately half a dozen manuals and booklets.

SPECIAL SERVICES: Consultation and technical information services are available only to Canadian inquirers.

B27

CANADIAN INSTITUTE OF MINING AND METALLURGY (CIM)
906-1117 Ste. Catherine St., W.
Montreal 110, Que., Canada

PURPOSE AND FUNCTION: A technical society established to promote the arts and sciences relating to economical production of valuable minerals and metals by means of technical meetings and publication of technical papers.

PUBLICATIONS:
Canadian Mining and Metallurgy Bulletin (monthly).
Canadian Metallurgical Quarterly. Sponsored by the Mines Branch, Department of Energy, Mines and Resources, in cooperation with CIM.
Transactions of the Canadian Institute of Mining and Metallurgy and of the Mining Society of Nova Scotia (annual).
Special volumes pertaining to the mineral industry (periodic).

B28

CANADIAN WELDING SOCIETY
73 Adelaide St., W.
Toronto, Ont., Canada

PURPOSE AND FUNCTION: To promote the art and science of welding through seminars, lectures, exhibits, and educational courses.

B29

CENTRE INTERNATIONAL DE DÉVELOPPEMENT DE L'ALUMINIUM (CIDA)
20, rue de la Baume
Paris, 8e, France

PURPOSE AND FUNCTION: A trade association for European aluminum industries that serves as the central point for the discussion and pursuit of coordinated technical policies and research and development projects.

PUBLICATIONS: *Aluminium Abstracts* (fortnightly, 1963–April 1970). Merged in 1970 with *Aluminum Technical Information Service* to form *World Aluminum Abstracts.* (*Aluminium Abstracts* was prepared for CIDA by the Aluminium Federation, London and Birmingham, England, which also prepares abstracts for *World Aluminum Abstracts.*) Preceded by *Light Metals Bulletin* (1939–1962). For description of *World Aluminum Abstracts*, see American Society for Metals (B15).

Also issues technical reports and data sheets which are submitted to the International Standards Organization (ISO) for consideration in preparing ISO aluminum recommendations.

B30

COPPER DEVELOPMENT ASSOCIATION (U.K.)
Orchard House, Mutton Lane
Potters Bar (Hertfordshire),
England

PURPOSE AND FUNCTION: To foster the application of copper and copper-base alloys through the dissemination of technical advice and information on the metal, its alloys, and compounds. CDA is one of the national copper development and information centers sponsored by copper producers and fabricators through the International Copper Development Council (CIDEC).

PUBLICATIONS:
Primary Journal: The Association provides material for *Copper* (quarterly). Published by CIDEC in English and four other languages.
Abstracts: Copper Abstracts (bi-monthly, 1959–), also published by CIDEC in five languages. Approximately 600 abstracts annually from more than 200 journals, as well as patents, books, reports, and pamphlets. Abstracts are arranged in broad subject categories; annual author and subject indexes.

Books: An existing series of textbooks on properties and applications of copper is being replaced by a series of technical notes on specific areas.

INFORMATION RETRIEVAL SERVICE: The CIDEC Information Service, operated by International Copper Development Council. More than 200 technical journals are scanned regularly, and about 2,000 bibliographic references are indexed annually on optical coincidence or "peek-a-boo" cards. Since January 1970, indexing is based on the *Thesaurus of Terms on Copper Technology*, prepared by the Copper Data Center at Battelle Memorial Institute (D8) for the Copper Development Association (N.Y.). Prior to 1970, a keyword indexing system was used. The information store contains about 40,000 items accumulated during the past 35 years. Search requests are filled by sending copies of appropriate abstracts, but the service essentially exists to provide personal replies to specific technical problems. There is no charge for this service, although inquiries are usually channeled through the various national copper development centers.

B31

COPPER DEVELOPMENT ASSOCIATION, INC. (CDA)
405 Lexington Ave.
New York, N.Y. 10017

PURPOSE AND FUNCTION: An association of U.S. copper mining companies, smelters, refiners, brass mills, wire and cable mills, and foundries. The Association seeks to expand the applications of copper, brass, and bronze in five major markets: transportation, building construction, electrical and electronics industries, industrial machinery and equipment, and consumer and general products.

PUBLICATIONS: *Standards Handbook.* Contains tolerances, alloy data for wrought and cast products, terminology, engineering data, sources, and specifications cross-index.
Copper Topics (quarterly).

Annual Data: Copper Supply & Consumption.

Application Data Sheets, Technical Reports, and product handbooks.

Publications booklet available on request. See Copper Data Center (D8) for abstracting and indexing publications.

TECHNICAL INFORMATION SERVICES: CDA sponsors the Copper Data Center (D8), a computer-based information retrieval system covering the world's technical information on copper and copper alloys, at Battelle Memorial Institute in Columbus, Ohio. The Center's technical services are available at no charge; telephone or mail inquiries may be addressed to CDA or to the Copper Data Center.

B32

DUCTILE IRON SOCIETY (DIS)
P.O. Box 858
Cleveland, Ohio 44122

PURPOSE AND FUNCTION: A trade association that conducts metallurgical research, performs product development and promotion, and sponsors research at engineering schools.

PUBLICATIONS:
Ductile Iron Data (quarterly).
Ductile Iron Technical Notes and News (quarterly).
Bulletins of papers presented at meetings (quarterly).

B33

ELECTROCHEMICAL SOCIETY (ECS)
30 E. 42nd St.
New York, N.Y. 10017

PURPOSE AND FUNCTION: A technical society devoted to advancing the theory and practice of electrochemistry, electrometallurgy, electrothermics, electronics, and allied subjects. Its objectives are to ensure research and its reportage and availability of adequate training. Six of its nine divisions cover various aspects of metallurgy: Corrosion; Electrodeposition; Electronics (including Semiconductors); Electro-

thermics and Metallurgy; Industrial Electrolytic; and Theoretical Electrochemistry.

PUBLICATIONS:
Primary Journals: Journal of the Electrochemical Society (monthly). Published as a single unit in four sections: *Electrochemical Science, Solid State Science, Electrochemical Society Reviews and News,* and *Electrochemical Technology.* Annual bound volume, including all four sections, also available.

Abstracts and Reviews: Extended Abstracts (semiannual). Softbound volume of condensed papers presented at spring and fall meetings.

The *Electrochemical Society Reviews and News* section of the *Journal of the Electrochemical Society* publishes invited and contributed reviews of current research.

Other: Monographs are published periodically on special subjects and symposia papers.

B34

FORGING INDUSTRY ASSOCIATION (FIA)
1121 Illuminating Bldg.
55 Public Sq.
Cleveland, Ohio 44113

PURPOSE AND FUNCTION: A trade association for manufacturers of closed-impression die forgings and manufacturers of equipment and supplies used by the forging industry. FIA supports the Forging Industry Educational and Research Foundation.

PUBLICATION: *Forging Topics* (quarterly).

B35

GRAY AND DUCTILE IRON FOUNDERS' SOCIETY (GDIFS)
National City-East Sixth Bldg.
Cleveland, Ohio 44114

PURPOSE AND FUNCTION: A trade association that provides statistical data, marketing information, technical services, and research for executives and owners of foundries providing gray and ductile iron castings. Metallurgical interests cover the iron-carbon-silicon system, metallurgy of cast iron, iron casting production, and engineering properties.

PUBLICATIONS:
Gray and Ductile Iron News (monthly).
The Iron Castings Handbook (latest edition, 1971).
Summary of Specifications for Iron Castings.

SPECIAL SERVICES: Consultation on the properties and applications of cast irons and casting design is available without charge.

B36

INDUSTRIAL HEATING EQUIPMENT ASSOCIATION (IHEA)
2000 K St., N.W.
Washington, D.C. 20006

PURPOSE AND FUNCTION: A trade association for manufacturers of industrial furnaces, combustion equipment, and dielectric heaters that deals primarily with industrial information, standards, and specifications.

LIBRARY SERVICES: Limited library services are available to members and government officials.

B37

INSTITUTE OF BRITISH FOUNDRYMEN (IBF)
137-139 Euston Rd.
London, N.W.1, England

PURPOSE AND FUNCTION: A technical society for the foundry industry.

PUBLICATIONS: *British Foundryman* (monthly).
Reference works such as *Typical Microstructures of Cast Iron*, 2nd Edition, 1966, and *Atlas of Defects in Castings*, 2nd Edition, 1961.
A number of individual conference papers. List available on request.

B38

INSTITUTE OF METAL FINISHING (IMF)
178 Goswell Rd.
London, E.C.1, England

PURPOSE AND FUNCTION: A technical society that promotes the study of

metal finishing, encourages research, and publishes and disseminates information on the science and technology of metal finishing.

PUBLICATIONS: *Transactions of the Institute of Metal Finishing* (five times per year).
Conference proceedings and symposia.

B39

INSTITUTE OF METALS (IOM)
(AND INSTITUTION OF METALLURGISTS)
17 Belgrave Sq.
London, S.W.1, England

PURPOSE AND FUNCTION: The Institute promotes the science and practice of metallurgy in all of its branches and facilitates the exchange of ideas among members and the community through meetings and publications. Originally, its interests were restricted to nonferrous metals; but, in recent years, many of its publications and services have been expanded to include ferrous interests, particularly in areas of basic, fundamental, and theoretical metallurgy. Principal functions of the Institution of Metallurgists are the establishment of professional standards, career development, and educational activities.

LIBRARY: The Institute and the Institution, together with the Iron and Steel Institute, cosponsor the Joint Metallurgical Library, which contains more than 30,000 books and reports and more than 2,000 periodical titles.

PUBLICATIONS:
Primary Journals:
Metals and Materials (monthly). Contains news and articles of general interest.
Journal of the Institute of Metals (monthly). Contains original papers dealing with all aspects of nonferrous metallurgy, including applied research and engineering.
Metal Science Journal (bimonthly). Contains original papers dealing with physical and chemical metallurgy, including ferrous and

nonferrous metals and other materials. The Iron and Steel Institute collaborates in editing this *Journal*.

Powder Metallurgy (semiannual). Official publication of the Powder Metallurgy Joint Group of the Iron and Steel Institute, the Institute of Metals, and the Institution of Metallurgists, devoted to the science and use of metal powders.

Abstracts, Indexes, and Reviews:

Metals Abstracts and *Metals Abstracts Index*, published jointly with the American Society for Metals (see B15).

Metallurgical Abstracts (1908–1967 inclusive). Predecessor publication to *Metals Abstracts*.

International Metallurgical Reviews (quarterly, March 1972–). Contains authoritative critical reviews of various aspects of metallurgical science and technology. Preceded by *Metallurgical Reviews* (1956–1971), published separately through 1966, then incorporated in *Metals and Materials* through 1971.

Books, Monographs, and Conference Proceedings: Approximately 30 titles appear in the current catalog (including some sponsored by the Institution of Metallurgists).

SPECIAL INFORMATION SERVICES: As joint publisher with the American Society for Metals of *Metals Abstracts* and *Metals Abstracts Index*, serves as leasing agent overseas for the METADEX computer tapes (see B15).

B40

INSTITUTION OF ELECTRICAL ENGINEERS (IEE)
INSPEC Marketing Department
P.O. Box 8, Southgate House
Stevenage, Herts., England

PURPOSE AND FUNCTION: A technical society serving the broad field of electrical engineering and electronics. Secondary publications and data bases cover limited metallurgical areas. IEE's information program is known as International Information Services in Physics, Electrotechnology, Computers, and Control (INSPEC).

PUBLICATIONS:
Abstracts and Indexes:
Science Abstracts. Published in three separate subscription journals: A. *Physics Abstracts;* B. *Electrical and Electronics Abstracts;* C. *Computer and Control Abstracts.* All are issued monthly. Sections A and B each contain approximately 30,000 to 36,000 abstracts annually, arranged in appropriate subject categories. A series of cumulated subject and author indexes is available.

Current Papers in Physics, Current Papers in Electrical and Electronics Engineering, and *Current Papers on Computers and Control* are companion publications which provide title, author, and bibliographic references only.

INFORMATION RETRIEVAL SERVICES: IEE conducts current awareness and retrospective searches of magnetic tape data files (INSPEC) generated for each of the three journals of *Science Abstracts.* The U.S. marketing agent for INSPEC tapes and search services is 3i Company/Information Interscience, Inc. (E19).

Topics (weekly). Consists of 4 × 6 cards containing information, selected by computer search of INSPEC files, on any of 21 subjects or topics. Several subjects are of metallurgical interest, such as "Thin Magnetic Films" and "Semiconductor Fabrication."

B41

INSTITUTION OF MINING AND METALLURGY (IMM)
44 Portland Pl.
London, W.1, England

PURPOSE AND FUNCTION: A technical society devoted to the science and practice of mining of minerals other than coal and with the metallurgy of metals other than iron. IMM operates its own library and information service.

PUBLICATIONS:
IMM Bulletin (monthly).
Transactions of the Institute (annual).
IMM Abstracts (monthly, 1950–).

A survey of world literature on economic geology, mining, mineral dressing, extractive metallurgy, and allied subjects; approximately 2,300 abstracts per year taken from nearly 600 journals and conference papers, arranged in sections according to the *Universal Decimal Classification*.

B42

INSTRUMENT SOCIETY OF AMERICA (ISA)
400 Stanwix St.
Pittsburgh, Pa. 15222

PURPOSE AND FUNCTION: A nonprofit scientific, technical, and educational organization dedicated to advancing the knowledge of, and practices related to, the theory, design, manufacture, and use of instruments and controls in science and industry. ISA is concerned with the application of all aspects of instrumentation to the industrial, laboratory, biophysical, marine, and space environments. Metallurgical instrumentation, in the form of process control instruments, temperature measurement, instruments used for physical testing of metals, and use of metals in instruments, comprises a small but important segment of ISA interest.

LIBRARY HOLDINGS: Publications of ISA and the International Federation of Automatic Control, plus those of other publishers and organizations with related interests.

PUBLICATIONS:
Primary Journals: Instrumentation Technology (monthly).
ISA Transactions (quarterly).
English translations of four Russian instrumentation journals, the most important, from a metallurgical standpoint, being *Industrial Laboratory*.
Books: Approximately 25 books, monographs, and reference works; plus an extensive set of *ISA Conference Proceedings*, including an annual conference on *Instrumentation for the Iron and Steel Industry*.
Abstracts, Indexes, and Bibliographies: Instrumentation Index is a reference retrieval service published in three sections: *Permuted Title Index, Bibliography*, and *Author Index*. It facilitates retrieval of data from all technical literature published by ISA—including proceedings of annual conferences and symposia—and articles appearing in the Society's journals—including those from the translated Russian journals. Published as frequently as necessary to cover all ISA literature as it appears. The first issue was published in 1967.
ISA Abstracts (annual). Booklet containing abstracts of all papers presented at ISA annual conferences.
Standards and Specifications: More than 30 publications, mostly nonmetallurgical, sold separately.

B43

INTERNATIONAL LEAD ZINC RESEARCH ORGANIZATION (ILZRO)
292 Madison Ave.
New York, N.Y. 10017

PURPOSE AND FUNCTION: A trade and research association that conducts fundamental and applied research to create new products, develops new uses for existing products, creates additional market outlets, and fosters research to compile new technical knowledge on the two metals. In 1970, ILZRO was conducting 60 research programs on lead and 45 programs on zinc.

LIBRARY HOLDINGS: Approximately 500 volumes.

LIBRARY SERVICES: Not open to the public. Provides services to member companies and industry interested in research; both direct and interlibrary loan to member companies only.

PUBLICATIONS: *ILZRO Research Digest* (semiannual). In five parts, published separately:
Part I—Die Cast and Wrought Zinc.
Part II—Zinc for Corrosion Protection.
Part III—Zinc Chemistry.
Part IV—Lead Metallurgy.
Part V—Lead Chemistry.

A small number of manuals and contractor reports.

Abstracts: Distributes *Lead Abstracts* and *Zinc Abstracts* (see ZDA/LDA Abstracting Service, F13).

B44

INTERNATIONAL MICROSTRUCTURAL ANALYSIS SOCIETY (IMS)
P.O. Box 219
Los Alamos, N.M. 87544

PURPOSE AND FUNCTION: A technical society for those engaged in the investigation and analysis of microstructures of materials such as metals, ceramics, cermets, minerals, plastics, chemicals, aerosols, petroleum, and petrochemicals using, among other techniques: optical metallography, autoradiography, microradiography, field ion and field emission microscopy, quantitative techniques, electron microprobe analysis, electron diffraction analysis, and electron microscopy (replica, transmission, and scanning).

PUBLICATIONS:
Primary Journals: Microstructures (bimonthly). Published by A. Z. Publishing Corp. for IMS.
Metallography (quarterly). Published by American Elsevier Publishing Co. in association with IMS.
An annual *Proceedings* volume.

B45

INVESTMENT CASTING INSTITUTE (ICI)
3525 W. Peterson Rd.
Chicago, Ill. 60645

PURPOSE AND FUNCTION: A trade association for the manufacturers of precision investment castings for industrial use.

PUBLICATION: *Investment Casting Handbook.*

B46

IRON AND STEEL INSTITUTE (ISI)
1, Carlton House Terr.
London, S.W.1, England

PURPOSE AND FUNCTION: A technical and professional society for those engaged in production or fabrication of iron or steel or involved with the application of iron and steel. ISI's interests include plant installation, industrial and management aspects, economics, and history.

LIBRARY HOLDINGS: Maintains, in cooperation with the Institute of Metals and the Institution of Metallurgists, the Joint Metallurgical Library, which contains more than 30,000 books and reports and more than 2,000 periodical titles.

PUBLICATIONS:
Primary Journals: Journal of the Iron and Steel Institute (monthly).
Steel in the USSR (until January 1971 known as *Stal' in English*) (monthly). A selected translation of the Russian journals *Stal'* and *Izvestiya VUZ Chernaya Metallurgiya.*
Books: More than 80 titles, including conference proceedings, meeting reports, and special reports. Catalog available on request.
Abstracts and Indexes: Titles of Current Literature (until January 1970 known as *Abstracts of Current Literature*) (monthly). Included as a section in the *Journal of the Iron and Steel Institute.* For details, see description of ABTICS below.
Annual Index to Publications of the Iron and Steel Institute contains author and subject indexes to all papers published in the *Journal,* bibliographies, ABTICS abstracts, and other ISI publications. Includes list of periodicals covered by ABTICS.
Bibliographies: A series on various technical topics currently numbering about 25 titles.
Other: British Iron and Steel Industry Translation Service (BISITS). A storehouse for quality translations by ISI and others in the steel industry in order to avoid duplication. Copies are made and offered for sale by ISI. Regular lists of new translations and those in process are circulated free to anyone interested; the list is also published in the *Journal of the Iron and Steel Institute.*

World Calendar of Forthcoming Meetings (bimonthly). Provides basic details about worldwide meetings concerned with both ferrous and non-ferrous metallurgy. Each issue contains three indexes—title, place, and operating body—and covers a period of about two years.

INFORMATION RETRIEVAL SERVICES: Abstract and Book Title Index Card Service (ABTICS). An alerting service and bibliographic search tool consisting of abstracts on 3 × 5 cards, delivered weekly, and book cards containing title, bibliographic information, and subject codes but no abstract, delivered monthly. The Service was started in 1960. The cards contain the same *Abstracts of Current Literature* as appeared in the *Journal* through 1969 and correspond to the *Titles of Current Literature* published since then. Approximately 9,000 abstracts are issued annually. Arrangement in the *Journal* is by 46 subject headings. In the card service, the abstracts carry *Universal Decimal Classification* (UDC) numbers for filing in subject order, both in a full and in a simplified, abbreviated system.

Iron and Steel Industry Profiles (ISIP) is a weekly Selective Dissemination of Information (SDI) service based on the ABTICS cards and BISITS translations. The cards and translation notices are organized into 20 profiles or subject interest categories, available by individual subscription. Delivery is made weekly in the form of profile sheets carrying abstracts, book titles, and titles of current translations falling within the subject scope of the profile. A catalog-brochure, with specimens and a finding index of more than 1,000 terms, is available free of charge.

SELF-HELPS: *Occasional Bulletins* are issued irregularly by the Library and Information Department. One of these, *Occasional Bulletin No. 3*, describes the simplified *UDC* schedule used on the ABTICS cards and provides an alphabetical subject index for the schedule.

Another publication, *Universal Decimal Classification—Special Subject Edition for Metallurgy*, ISI Special Report No. 4, 1964, 165 pp., gives the complete schedules of all the *UDC* classes dealing with metallurgy. It also includes an alphabetical subject index.

Another *Occasional Bulletin*, issued annually, lists the periodicals abstracted in ABTICS.

B47

LEAD INDUSTRIES ASSOCIATION (LIA)
292 Madison Ave.
New York, N.Y. 10017

PURPOSE AND FUNCTION: A trade association of mining companies, smelters, refiners, and manufacturers of lead products and components.

PUBLICATIONS: *Lead* (quarterly).

Literature of technical and semi-technical nature, including standards and specifications.

B48

MALLEABLE FOUNDERS' SOCIETY (MFS)
781 Union Commerce Bldg.
Cleveland, Ohio 44115

PURPOSE AND FUNCTION: A trade association for manufacturers of malleable iron castings. Administers the Malleable Research and Development Foundation.

PUBLICATIONS: *Malleable Iron Castings Handbook.*
Occasional books and brochures.

B49

MALLEABLE RESEARCH AND DEVELOPMENT FOUNDATION (MRDF)
781 Union Commerce Bldg.
Cleveland, Ohio 44115

PUBLICATIONS: *Modern Pearlitic Malleable Castings Handbook.*
Approximately a dozen special reports and bulletins.

B50

METAL POWDER INDUSTRIES FEDERATION (MPIF)
and
AMERICAN POWDER METALLURGY INSTITUTE (APMI)

201 E. 42nd St.
New York, N.Y. 10017

PURPOSE AND FUNCTION: A federation of five trade associations: Powder Metallurgy Parts Association, Metal Powder Producers Association, In-Plant Powder Metallurgy Association, Powder Metallurgy Equipment Association, and Magnetic Powder Core Association; plus a technical society, the American Powder Metallurgy Institute. Purposes of the Federation are to advance and promote metal powders and the products of powder metallurgy through investigation, research, and the exchange of ideas.

LIBRARY HOLDINGS: A complete collection of powder metallurgy books and periodicals.

LIBRARY SERVICES: Open to the public for reference; interlibrary loan and photocopy services.

PUBLICATIONS:
Primary Journals: International Journal of Powder Metallurgy (quarterly).
Powder Metallurgy Information Bulletin (monthly).
Books: Current catalog contains 32 titles, including *Handbook of Metal Powders* (published by Reinhold) and a number of annual conference proceedings appearing under the title *Progress in Powder Metallurgy.*
Bibliographies and Reviews: Three bibliographies: *Metal Powder Technology, Powder Metallurgy Parts,* and *Cermets and Cemented Carbides* (prepared by the American Society for Metals). The *Journal* includes critical reviews of pertinent literature.
Other: Numerous standards and specifications, available either separately or as complete sets in binders.

B51

METAL TREATING INSTITUTE (MTI)
Box 448
Rye, N.Y. 10580

PURPOSE AND FUNCTION: A national trade association of commercial heat treating companies.

PUBLICATIONS: *Guide to Heat Treating Services.*
Two programmed learning texts covering the principles and theory of heat treating.

B52

THE METALLURGICAL SOCIETY (TMS)
AMERICAN INSTITUTE OF MINING,
METALLURGICAL AND PETROLEUM
ENGINEERS (AIME)
345 E. 47th St.
New York, N.Y. 10017

PURPOSE AND FUNCTION: A professional engineering society devoted to advancing the arts and sciences by which mankind utilizes the metallic elements of the earth's resources. Through its publications, conferences, and local activities, TMS encourages the exchange of information and ideas. One of three constituent societies of AIME, TMS has three divisions: Extractive Metallurgy, Institute of Metals, and Iron and Steel.

LIBRARY: One of the supporting members of the Engineering Societies Library (A6), housed in the same building.

PUBLICATIONS:
Primary Journals: Journal of Metals (monthly).
Metallurgical Transactions (monthly). Published jointly with American Society for Metals.
Journal of Electronic Materials (quarterly).
Books: Current catalog lists 42 titles, primarily proceedings of meetings, conferences, and symposia.
Abstracts: Collections of abstracts of recent spring and fall meetings.
Other: The *TMS Paper Selection Program.* Separate copies of selected papers presented at spring and fall meetings; nearly 200 titles listed.

B53

NATIONAL ASSOCIATION OF CORROSION ENGINEERS (NACE)
2400 W. Loop S.
Houston, Tex. 77027

PURPOSE AND FUNCTION: A nonprofit

scientific and research association concerned with the reactions of all materials to corrosive environments and with corrosion control measures.

LIBRARY SERVICES: A small specialized library open to the public for reference; photocopy service.

PUBLICATIONS:
Primary Journals: Corrosion (monthly).
Materials Protection and Performance (monthly).
Books: Conference proceedings and miscellaneous books.
Abstracts: Corrosion Abstracts (bimonthly, (1962–). Approximately 4,000 abstracts annually from about 1,000 journals, and books, pamphlets, and reports. Subject index in each issue. An annual bound volume, *Corrosion Abstracts Yearbook*, contains cumulated subject and author indexes. Preceded by *Bibliographic Surveys of Corrosion,* 1945–60.
Other: Numerous standards, specifications, and special reports.

INFORMATION RETRIEVAL SERVICE: Conducts literature searches at an hourly fee.

SELF-HELPS: An *Abstract Filing System Classification* has been developed by the NACE Abstract Committee and is used for arrangement of *Corrosion Abstracts.* Topical index headings used in the system are published in *Corrosion Abstracts Yearbook.*

B54

NON-FERROUS FOUNDERS' SOCIETY (NFFS)
21010 Center Ridge Rd.
Cleveland, Ohio 44116.

PURPOSE AND FUNCTION: A trade association serving manufacturers of brass, bronze, aluminum, and other nonferrous castings.

PUBLICATIONS: Handbooks on *Copper, Brass and Bronze, Their Structures, Properties and Applications;* and *Design of Sand and Permanent Mold Aluminum Castings.*
A number of other books, special

reports, and pamphlets, primarily on industrial and management topics.

B55

OPEN DIE FORGING INSTITUTE (ODFI)
440 Sherwood Rd.
La Grange Park, Ill. 60525

PURPOSE AND FUNCTION: A trade association to provide statistics, promotional information, and related technical materials covering specifications and standards for manufacturers of iron and steel forgings produced on open dies other than drop forgings.

PUBLICATIONS:
Open Die Forging Manual.
Allowances and Tolerances Chart.
Tabulations of Frequently Used Specifications for Steel Forgings.

B56

PRODUCTION ENGINEERING RESEARCH ASSOCIATION (PERA)
Melton Mowbray
Leicestershire, England

PURPOSE AND FUNCTION: PERA research programs, in general, embrace the design, performance, and utilization of machine tools, electrochemical and conventional machining techniques, explosive forming, other metal forming techniques, and automation. It operates an advanced information service and library covering conventional and nonconventional metal-forming techniques and the design and utilization of machine tools.

PUBLICATION: *PERA Bulletin* (monthly, 1947–). Contains short, indicative abstracts of articles dealing with forming of metals, foundry practice, inspection, machinery, materials handling, mechanical assembly, plating, painting and spraying, powder metallurgy, power and power transmission, products and their manufacture, welding, brazing, soldering, and flame cutting. It covers approximately 500 periodicals, plus standards and patents, for both ferrous and nonferrous metals.

Issues are not cumulated, and there is no index.

SPECIAL INFORMATION SERVICES: These vary from making an evaluated state-of-the-art report following an exhaustive literature survey of a particular subject to simply making a list of all relevant material without assessing its pertinence. Patent and product surveys of catalogs and other trade literature also can be made.

The Express Monitoring Service offers a tailor-made weekly, fortnightly, or monthly bulletin of abstracts and reports on any commercial or technical subject covered by the Association.

B57

SELENIUM-TELLURIUM DEVELOPMENT ASSOCIATION, INC. (STDA)
475 Steamboat Rd.
Greenwich, Conn. 06830

PURPOSE AND FUNCTION: An association formed by the primary producers of selenium and tellurium to stimulate interest in these two elements and to promote improved applications and new uses, primarily through sponsored research.

LIBRARY SERVICES: Collects and makes available copies of publications and reports dealing with selenium and tellurium in various aspects of chemistry, metallurgy, nutrition, physics, and electronics.

PUBLICATIONS: *Bulletin of the STDA* (irregular). Contains news of the industry and announcements of sponsored research, meetings, and publications.

Abstracts: Selenium and Tellurium Abstracts (monthly, 1955–). Prepared for the Association by Chemical Abstracts Service. Approximately 3,000 abstracts per year from 13,000 journals, books, reports, conference proceedings, and patents. Annual subject and author indexes. Free on request.

B58

SOCIETY OF AEROSPACE MATERIAL AND PROCESS ENGINEERS (SAMPE)

P.O. Box 613
Azusa, Calif. 91702

PURPOSE AND FUNCTION: A technical society that advances aerospace material and processing technology through technical communications (publications and conferences), education, and recognition of the profession. Covers all aspects of advanced metallurgy, oriented to aerospace, including coatings, processing, welding, properties, and applications.

PUBLICATIONS:
Primary Journals: SAMPE Journal (monthly); *SAMPE Quarterly.*
Books: About two volumes per year of symposia in the *National SAMPE Technical Conference Series.*

B59

SOCIETY FOR ANALYTICAL CHEMISTRY (SAC)
9/10 Savile Row
London, W.1, England

PURPOSE AND FUNCTION: A learned society dealing with all branches of analytical chemistry.

PUBLICATIONS:
The Analyst (monthly).
Proceedings of the Society for Analytical Chemistry (monthly).
Occasional books.
Abstracts: Analytical Abstracts (monthly, 1954–). Approximately 9,500 abstracts per year taken from nearly 600 journals, plus reports, conference proceedings, and patents. Arranged in nine main subject categories, three of which are of some metallurgical interest: General Analytical Chemistry, Inorganic Chemistry, and Techniques and Apparatus. Semiannual subject and author indexes, published separately, contain a list of publications abstracted.

B60

SOCIETY OF AUTOMOTIVE ENGINEERS (SAE)
Two Pennsylvania Plaza
New York, N.Y. 10001

PURPOSE AND FUNCTION: A professional society of engineers in the field of self-propelled ground, flight, and space vehicles, its purpose is to promote the arts, sciences, standards, and engineering practices related to such mechanisms and associated equipment. The metallurgical interests are of considerable importance, particularly in the areas of standard alloy compositions and heat treatments. The history of SAE standards for steel composition dates back to 1910, and the numbering system has since been closely correlated with that of the American Iron and Steel Institute. These specifications are updated annually in the *SAE Handbook*, as are the specifications for nonferrous materials used in the automotive industry. SAE is also responsible for *Aerospace Materials Specifications* (AMS).

LIBRARY: The Engineering Societies Library (A6) is used as a depository.

PUBLICATIONS: Approximately 1,000 new titles are issued each year.

Primary Journals: Automotive Engineering (monthly). (Also known as *SAE Journal of Automotive Engineering*.)

SAE Transactions (annual).

Books: Conference proceedings, special publications, and a number of volumes each year in the *Advances in Engineering* series and the *Progress in Technology* series.

SAE Handbook (annual).

Abstracts and Indexes: The "New Publications" section in each issue of *Automotive Engineering* provides abstracts of SAE technical papers and reports, together with an order card for purchasing at a unit price.

Transactions Abstract-Index (see *Transactions*, Plan D, below).

Index of Aerospace Materials Specifications. A 26-page booklet listing chemical compositions with cross-references to similar specifications and materials.

Cumulative Subject/Author Index, 1965–1970. Covers all technical papers presented before the Society during this period.

Other: Technical papers (more than 700 per year) issued separately

and distributed in various forms and by various means; technical reports including standards, recommended practices, information reports, research reports, and *Handbook* supplements.

Standards and specifications for automotive materials are included in the *SAE Handbook*. The *Aerospace Materials Specifications* are available either as separate specifications or in collected sets.

SPECIAL INFORMATION SERVICES: The "New Publications" section of *Automotive Engineering* is an alerting service.

The *SAE Transactions* is offered under four annual plans: Plan A—a library edition containing more than 200 selected, full-length, technical papers; Plan B—a microfiche version of the library edition; Plan C—SAE Technical Literature on Microfiche, the entire yearly technical literature output, consisting of 700 papers; Plan D—*Transactions Abstract-Index*, a complete subject and author index covering all papers presented before the Society.

B61

SOCIETY OF CHEMICAL INDUSTRY (SCI)
14 Belgrave Sq.
London, S.W.1, England

PURPOSE AND FUNCTION: A technical society founded to provide interchange of ideas with respect to improvements in the various chemical industries, discussion of all matters bearing upon the practice of applied chemistry, and publication of information thereon.

PUBLICATIONS:
Chemistry and Industry (weekly).
Journal of Applied Chemistry (monthly, 1950–). Contains an "Abstracts" section which provides approximately 8,000 abstracts per year from worldwide journals, books, and patents. Arranged in seven major subject classes; one of interest is "Industrial Inorganic Chemistry," containing subsections on General Metallurgy, Ferrous Metallurgy

—Iron, Ferrous Metallurgy—Steel, Nonferrous Metallurgy, Electrochemical, and Corrosion. An author index appears in each issue.

Miscellaneous books and monographs.

B62

SOCIETY OF DIE CASTING ENGINEERS (SDCE)
14530 W. Eight Mile Rd.
Detroit, Mich. 48237

PURPOSE AND FUNCTION: A technical society dedicated to increasing the knowledge of die casting; fostering educational programs; encouraging the investigation of new techniques; and advancing, through education and research, international development and growth in the die casting and allied industries. Interests cover nonferrous alloys used in die casting, steels and other materials used for dies, melting, and treatment of molten metals.

PUBLICATIONS:
Primary Journal: Die Casting Engineer (monthly).
Books: Die Casting Congress Transactions; a handbook in preparation.

SPECIAL SERVICES: Consultation on die casting.

B63

SOCIETY FOR EXPERIMENTAL STRESS ANALYSIS (SESA)
21 Bridge Sq.
Westport, Conn. 06880

PURPOSE AND FUNCTION: A technical society covering the measurement of stresses and strains as applied to metals and other materials.

PUBLICATIONS:
Primary Journals: Journal of Experimental Mechanics (monthly).
SESA Proceedings (semiannual).
Books: Handbook of Experimental Stress Analysis.
Indexes: Cumulative Indexes for SESA Proceedings.

B64

SOCIETY OF MANUFACTURING ENGINEERS (SME)

20501 Ford Rd.
Dearborn, Mich. 48126
(Formerly AMERICAN SOCIETY OF TOOL AND MANUFACTURING ENGINEERS)

PURPOSE AND FUNCTION: A professional society of tool and manufacturing engineers and management executives concerned with manufacturing techniques. Its purpose is to advance scientific knowledge in the field of manufacturing engineering and to apply the Society's resources to research, writing, publishing, and disseminating such information. Metallurgical interests are evident in six of its nine technical divisions: Casting, Molding, and Metallurgical Processing; Engineering Materials; Finishing and Coating; Inspection and Quality Control; Material Forming; and Material Removal.

LIBRARY HOLDINGS: More than 1,300 books; 200 journals.

LIBRARY SERVICES: Open to qualified scholars and students by appointment only; interlibrary loan.

PUBLICATIONS:
Primary Journals: Manufacturing Engineering and Management (monthly).
Technical Divisions Newsletter (monthly).
Books: The current book catalog contains 27 titles, plus 13 handbooks and instruction manuals. The *Tool Engineers Handbook*, published by McGraw-Hill, contains much information of metallurgical interest.
Abstracts, Indexes, and Bibliographies: SME Technical Digest contains abstracts of, and ordering information for, technical papers published by the Society. Issued annually through 1969 and quarterly beginning in 1970. Contains about 500 abstracts per year. Subject index in each issue.
Other: SME Technical Papers are available separately in either hard copy or on microfiche. They include papers presented at SME conferences and seminars, and those published in the Society's periodicals.

SELF-HELPS: Programmed learning

courses are offered for home study, and many clinics, seminars, and workshops are sponsored.

B65
SOCIETY OF MINING ENGINEERS (SME)
AMERICAN INSTITUTE OF MINING, METALLURGICAL AND PETROLEUM ENGINEERS (AIME)
345 E. 47th St.
New York, N.Y. 10017

PURPOSE AND FUNCTION: A professional society for those engaged in the finding, exploitation, treatment, and marketing of all classes of minerals. Its four divisions are: Mining and Exploration, Coal, Minerals Beneficiation, and Industrial Minerals. SME is one of three constituent societies of AIME.

LIBRARY: One of the supporting members of the Engineering Societies Library (A6), housed in the same building.

PUBLICATIONS:
Mining Engineering (monthly).
Transactions of SME/AIME (quarterly).

B66
STEEL CASTINGS RESEARCH AND TRADE ASSOCIATION (SCRATA)
East Bank Rd.
Sheffield, 3, England

PURPOSE AND FUNCTION: A research and trade association which provides library and information services covering the design, manufacture, and applications of steel castings, as well as subjects such as industrial health and safety in steel foundries.

PUBLICATIONS: *Steel Casting Abstracts* (bimonthly, 1952–). Approximately 1,500 abstracts per year covering about 250 journals, plus books, reports, conference proceedings, and pamphlets; annual subject and author indexes. Available to members gratis and to others on paid subscription.

SCRATA also publishes data sheets, recommended procedures, bibliographies, and reports. A useful reference book is *British and For-*

eign Specifications for Steel Castings, 3rd Edition, 1968, 118 pp.

B67
STEEL FOUNDERS' SOCIETY OF AMERICA (SFSA)
Westview Towers
21010 Center Ridge Rd.
Rocky River, Ohio 44116

PURPOSE AND FUNCTION: A trade association serving the manufacturers of carbon and high alloy castings that covers all aspects of production and use and sponsors research. The Alloy Casting Institute has recently become a division of SFSA. Its major objective is to broaden the application of cast high alloys through development of information beneficial to current and potential users. It has developed data that form the basis for high-alloy castings specifications.

PUBLICATIONS:
Journal of Steel Castings Research (quarterly).
Casteel Magazine (semiannual).
8 Plus Magazine (semiannual).
Steel Castings Handbook.
Above publications are available for purchase. Separate research reports are available to members only.

B68
TIN RESEARCH INSTITUTE (TRI)
483 W. Sixth Ave.
Columbus, Ohio 43201

PURPOSE AND FUNCTION: The U.S. affiliate of the Tin Research Institute in the United Kingdom, which is headquarters and research laboratory for the International Tin Research Council, financed by the major tin-ore-producing countries of the world. Originally sponsored at Battelle Memorial Institute, it is now an independent organization with its own building in the Battelle complex. The staff, however, has access to many of the Battelle facilities, including the chemical laboratories and a number of services. TRI's purpose is to develop the consumption of tin, to develop new uses, and to improve existing tin products and the processes by which they are made and used. TRI functions are: (1) to collect the results of research in the use of tin from the open literature and from laboratory studies; (2) to summarize and publish these results in trade journals; (3) to distribute informational manuals on specific tin uses; and (4) to provide a consulting service for tin users in the United States, Canada, and Mexico. Subject coverage includes properties of tin and its alloys, their fabrication and uses, as well as tinplate, solders, bronzes, white metals, tin and tin alloy plated coatings, and inorganic and organotin compounds.

LIBRARY AND LIBRARY SERVICES: Resources of the Battelle Columbus library are available for reference; direct and interlibrary loan and limited photocopy service.

PUBLICATIONS: *Tin and Its Uses* (quarterly).

Annual Report of the International Tin Research Council.

More than 450 miscellaneous technical publications.

SPECIAL INFORMATION SERVICES: Provides a technical advisory service which includes literature searches and consultation at either the Columbus Laboratories or in the user's plant. Maintains a card file of abstracts related to tin.

B69

WELDED STEEL TUBE INSTITUTE (WSTI)
522 Westgate Tower
Cleveland, Ohio 44126

PURPOSE AND FUNCTION: A trade association for manufacturers of tubes formed and welded from flat rolled carbon or stainless steel. Disseminates information to tubing users and encourages research and development.

PUBLICATIONS:
Tubular Steel Progress (three times per year).
Handbook of Welded Steel Tubing.

B70

THE WELDING INSTITUTE
Abington Hall
Abington, Cambridge, England

PURPOSE AND FUNCTION: A technical society dedicated to improving welding processes, operations, and results and to disseminating information thereon through meetings and forums, educational programs, and publications.

PUBLICATIONS:
Metal Construction and British Welding Journal (monthly).
Handbook of Welding Design, Vol. 1, 1967.
Multilingual Collection of Terms for Welding and Allied Processes, in four volumes: *General Terms; Arc Welding; Resistance Welding; Thermal Cutting.*

Symposia and proceedings volumes.

Sponsors English translations of two welding journals: *Automatic Welding (Avtomaticheskaya Svarka)* and *Welding Production (Svarochnoe Proizvodstvo)*.

SPECIAL INFORMATION SERVICES: An information storage and retrieval service is in the planning stage. Technical and trade information is to be available on rapid recall against specific subject inquiries. A current awareness (SDI) service also is planned.

B71

WIRE ASSOCIATION, INC. (WA)
209 Montowese St.
Branford, Conn. 06405

PURPOSE AND FUNCTION: A professional society devoted to increasing the technical skills and knowledge of those in the worldwide ferrous, nonferrous, and electrical wire industry.

PUBLICATIONS:
Wire Journal (monthly).
Steel Wire Handbooks. Vols. 1 and 2 deal with manufacture and processing of wire; Vol. 3, in preparation, will cover wire fabrication.

B72

ZINC INSTITUTE (ZI)
292 Madison Ave.
New York, N.Y. 10017

PURPOSE AND FUNCTION: A trade association serving the zinc industry of the United States and Canada. Its purposes are to disseminate information on the uses and characteristics of zinc and zinc products, to support and conduct studies and research on uses, and to collect and distribute marketing information.

PUBLICATIONS: The current catalog contains more than 100 titles, many of them promotional and statistical in nature, but also many of metallurgical interest in the areas of die casting, galvanizing, cathodic protection, and painting.

Abstracts: Distributes *Zinc Abstracts* (see ZDA/LDA Abstracting Service, F13) prepared in cooperation with the Zinc Development Association in England.

C

GOVERNMENT AGENCIES

ATOMIC ENERGY OF CANADA LTD.
(AECL)
Technical Information Branch
Chalk River Nuclear Laboratories
Chalk River, Ont., Canada

PURPOSE AND FUNCTION: AECL is a Crown company owned by and responsible to the Government of Canada. Its main activities are nuclear research and development, development and design of nuclear power systems, and production of radioactive isotopes and associated equipment. The Technical Information Branch is responsible for library and information services and for the distribution and sale of AECL reports and publications. Metallurgical interests include reactor component materials (corrosion, radiation effects), rare metals (composition, analysis, fabrication), and isotope devices used in metal rolling.

LIBRARY HOLDINGS: Approximately 90,000 books; nearly 3,000 periodicals; 228,000 reports.

LIBRARY SERVICES: Open for reference by letter appointment only; interlibrary loan and photocopy services; literature searches, restricted to nuclear disciplines.

PUBLICATIONS: Thousands of reports, reprints of journal articles, and translations. These are listed in *Publications of AECL* (annual, 1966–), plus supplements and a cumulated list for the years 1959–1966. These lists include a section on "Metals, Ceramics, and Other Materials," and each issue contains an author index, indexes of AECL and report numbers, and a list of depository libraries throughout the world.

C2

DEFENSE DOCUMENTATION CENTER
(DDC)
Cameron Station
Alexandria, Va. 22314

PURPOSE AND FUNCTION: DDC is sponsored by the Defense Supply Agency, Department of Defense (DOD), to support defense-related research, development, test, and evaluation activities by saving technical talent, time, and money through prevention of unnecessary duplication. It is the central facility within DOD for processing and supplying scientific and technical reports of DOD-sponsored activities. DDC is also a development and design center for DOD technical communication and information systems, services, and products.

RESOURCES: Copies of all reports

issued by DOD facilities and their contractors. These reports are identified by an Accession Document (AD) serial number. Currently, there are more than 670,000 titles, classified according to the *COSATI Subject Category List* (see National Technical Information Service SELF-HELPS, C10). Copies of reports are available in both hard copy and microfiche.

PUBLICATIONS:

Abstracts, Indexes, and Bibliographies: Technical Abstract Bulletin (TAB) (semimonthly). Announces latest report accessions in the "classified" and "limited-distribution" categories; approximately 1,500 per month. (For unclassified report announcements, see *Government Reports Announcements*, available from the National Technical Information Service, C10.) A companion index volume, also published semimonthly, is cumulated quarterly and annually. *TAB* and *TAB Index* are classified "confidential."

There are three types of bibliographies — Report Bibliographies, Scheduled Bibliographies, and Rapid Response Bibliographies. The Report Bibliography is tailored in response to a specific request; the Scheduled Bibliography follows a search technique of gathering references concerning subject areas of known or anticipated high demand; the Rapid Response Bibliography is merely a list of pertinent AD numbers transmitted by Telex.

The *Bibliography of Bibliographies* is an 8-volume compendium of more than 2,000 references to bibliographies that have been submitted to the DDC report collection.

INFORMATION RETRIEVAL SERVICES: Demand searches, which result in the Report Bibliographies noted above, are computer-generated.

SPECIAL SERVICES: The Research and Technology Work Unit Information System is maintained to provide quickly, information on ongoing research. The information is computer-stored to permit retrieval in a wide variety of logical combinations of the data elements. As of 1970, there were more than 20,000 active DOD work unit descriptions in the system. A search is requested by subject matter.

Defense R&D of the 1960's consists of descriptions, recorded on microfilm, of more than 400,000 scientific and technical documents accessioned by DDC during the 1960's. The reports are in accession number order, and the data packages include six indexes. The films are sold in four sections: two of unclassified documents available to the public and two of classified or limited-distribution documents.

DDC Referral Service directs authorized users to organizations or individuals who can provide information exceeding that contained in technical reports. A *Directory of the Defense Documentation Center Referral Data Bank* lists 182 centers operated for, or supported by, DOD or other federal agencies (see G3).

SELF-HELPS: DDC is responsible for the development and maintenance of the DOD *Thesaurus of Engineering and Scientific Terms* (TEST). This *Thesaurus* was prepared by the Office of Naval Research in cooperation with the Engineers Joint Council in New York and published in 1967.

USER QUALIFICATIONS: With a few exceptions, as noted, only registered DDC users may use the services of this facility. Eligible to become users are U.S. Government departments and agencies and their contractors, subcontractors, and grantees; potential defense contractors may also be eligible through Potential Defense Contractors Programs established by the military departments. A brochure entitled *Registration for Scientific and Technical Information Services of the Department of Defense*, DSAM 4185.3, gives detailed procedures for user registration in each of these categories. DOD reports that have no security or distribution limitations are available from NTIS (see C10).

C3

JOINT PUBLICATIONS RESEARCH
SERVICE (JPRS)
National Technical Information
Service
1000 N. Glebe Rd.
Arlington, Va. 22201

PURPOSE AND FUNCTION: A component of the National Technical Information Service (C10), JPRS was established to provide translation support for the U.S. Government. It selects and translates research and development literature from throughout the world, with emphasis on USSR and East European publications.

TRANSLATION SERVICE: Translations can be ordered individually or by standing order. The standing order service requires a deposit account with NTIS. For USSR translations, a standing order service is available for the subject catetory of "Materials, Metallurgy." Translations of documents from other countries do not have subject divisions. For a guide and index to the JPRS translations, see *Transdex* (CCM Information Corporation, E3).

PUBLICATIONS: JPRS publishes a series of abstract journals, two of which deal with metals: *USSR Scientific Abstracts, Materials Science and Metallurgy* (about 12 issues per year, 1966–), arranged in 27 subject categories; *East European Scientific Abstracts, Materials Science and Metallurgy* (about 8 issues per year, 1964–), arranged in 18 subject categories.

Photocopies of the original foreign language documents abstracted are available from the Library of Congress Photoduplication Service, Washington, D.C. 20540, for a fee.

C4

MINES BRANCH, DEPARTMENT OF
ENERGY, MINES AND RESOURCES
555 Booth St.
Ottawa, Ont., Canada

PURPOSE AND FUNCTION: The Mines Branch acts as the focal point in implementing the government's policy of providing scientific and technical support to Canada's mineral industry. Metallurgical interests include mineral beneficiation, pyrometallurgy, hydrometallurgy, environmental improvement, and physical metallurgy (melting, casting, solidification, forming and fabrication, alloying, corrosion, welding, and nondestructive testing).

LIBRARY HOLDINGS: Approximately 90,000 books and 2,500 serials.

LIBRARY SERVICES: Open to the public for reference; interlibrary loan and photocopy services.

PUBLICATIONS:
Mines Memo (Annual Review of Research Investigations of the Mines Branch).

Monographs, technical bulletins, information circulars, and research reports. One information circular annually abstracts *Scientific and Technical Papers Published by the Staff*.

INFORMATION RETRIEVAL: Preliminary steps have been taken to establish a scientific and technical information center.

C5

NATIONAL AERONAUTICS AND SPACE
ADMINISTRATION (NASA)
Scientific and Technical Information
Division, Office of Technology
Utilization
Washington, D.C. 20546

PURPOSE AND FUNCTION: The NASA Scientific and Technical Information Division assembles, summarizes, indexes, and stores the results of worldwide aerospace research and development activities. Facilities and services are more fully described in *The NASA Scientific and Technical Information System—and How to Use It*, available on request.

The NASA Office of Technology Utilization answers inquiries and sponsors Regional Dissemination Centers which provide computerized searching services to industry on a fee basis. Separate entries are provided for the Regional Dissemination Centers as follows:

Aerospace Research Applications Center (D2)

Knowledge Availability Systems Center (D16)

New England Research Application Center (D21)

North Carolina Science and Technology Research Center (D23)

Technology Application Center (D27)

Technology Use Studies Center (D28)

Western Research Application Center (D34).

METALLURGICAL COVERAGE: NASA's total subject coverage is fully described in a booklet, *The NASA Scientific and Technical Information System—Its Scope and. Coverage,* available on request. Descriptions are provided under 34 categories. The following is a quotation from Category 17, "Materials, Metallic":

> Exhaustive Interest—physical properties, metallurgy, testing and other aspects of metals, alloys and metallic compositions having demonstrated direct application to aircraft, space vehicles, rockets, launch vehicles, supporting facilities and other devices such as antennas and telescopes. Selective Interest—physical properties, applications, metallurgy, testing and evaluation of any metals, alloys and metal compositions having possible use in aerospace research or applications; methods of improving the properties of metals and alloys . . . and unusual properties and studies under extreme conditions.

Machining and forming processes are referenced in Category 15, "Machine Elements and Processes"; metals and alloys involved in nuclear applications are referenced in Category 22, "Nuclear Engineering." The booklet also lists specific metals, alloys, and properties that are covered.

RESOURCES: The NASA information bank contains more than half a million documents; thousands are added every month. These documents are from journal articles, reviews, and government, industry, research institute, and university reports which contain the details of findings by NASA personnel, contractors, subcontractors, and grant holders. These documents are abstracted and indexed for use in various publications and services. Bibliographic records and index terms are recorded on computer tape for retrieval services.

LIBRARY SERVICES: Eleven NASA-funded technical libraries operate on differing bases with varying levels of reference service; direct and interlibrary loan; photocopy service; and literature search capabilities. They are as follows:

Carnegie Library of Pittsburgh (A5)

Columbia University, New York

John Crerar Library (A7)

Georgia Institute of Technology, Atlanta, Ga.

Library of Congress (A8)

Linda Hall Library (A9)

Massachusetts Institute of Technology, Cambridge, Mass.

Southern Methodist University, Dallas, Texas

University of California Library, Berkeley, Calif.

University of Colorado Libraries, Boulder, Colo.

University of Washington Library, Seattle, Wash.

Representative collections of NASA publications are also maintained at a number of public libraries in major cities.

PUBLICATIONS:

Abstracts, Indexes, and Bibliographies: Scientific and Technical Aerospace Reports (STAR) (semimonthly, 1958–). Approximately 30,000 abstracts per year in 34 subject categories; metallurgical scope previously noted. Subject, author, corporate source, report number, and accession number indexes in each issue; cumulative indexes quarterly and semiannually. Available from the Superintendent of Documents, U.S. Government Printing Office (GPO), Washington, D.C. 20402. No periodical literature is abstracted in *STAR*. The related journal literature is covered in *International Aerospace Abstracts* (see B10).

Aeronautical Engineering—A Special Bibliography With Indexes (monthly, with annual cumulations, 1971–). Available from the National Technical Information Service, C10. A journal containing abstracts and indexes of the engineering and theoretical aspects of design, construction, evaluation, testing, operation, and performance of aircraft and associated components.

NASA Research and Technology Operating Plan Summaries (annual, 1970–). Available from NTIS as NASA TM-X- . (Formerly *NASA Research and Technology Program Flash Index*, 1964–69.) Presents technical summaries of the objectives of each program used for management review and control of research currently in progress; 14 groups, including 1 on "Physics, Materials, and Applied Mathematics Research."

Other: The following publications are available from NTIS:

NASA Tech Briefs (irregular). Available in 10 subject sections, priced separately. Of metallurgical interest are "Materials/Chemistry," "Physical Sciences," and "Fabrication Technology." These are 1- or 2-page industry-oriented summaries of innovations developed under NASA-sponsored scientific and technical programs. The series also includes *AEC-NASA Tech Briefs*. An index is available.

Technical Reports (NASA TR-R-). Scientific and technical information considered important, complete, and a lasting contribution to existing knowledge.

Technical Notes (NASA TN-D-). Information of lesser scope, but still important as a contribution to knowledge.

Technical Memoranda (NASA TM-X-). Information which receives limited distribution because of its preliminary or classified nature.

Technical Translations (NASA TT-F-). Information originally published in a foreign language, but considered sufficiently valuable to NASA's work to merit distribution in English.

Contractor Reports (NASA CR-). Technical information generated in the course of a NASA contract and released under NASA auspices.

Special Publications (NASA SP-). Conference proceedings, monographs, literature surveys, and continuing bibliographies. This series also includes *Technology Utilization Reports* and *Technology Utilization Surveys:* the former give detailed descriptions of developments and innovations of high promise; the latter are comprehensive state-of-the-art accounts identifying substantial contributions to technology. An annual catalog, *NASA Special Publications Currently Available*, also is offered.

INFORMATION RETRIEVAL SERVICES: Selected Current Aerospace Notices (SCAN), free to qualified NASA contractors only. A current awareness program established to inform scientists and engineers of recent articles and reports in their subject areas that are announced in *Scientific and Technical Aerospace Reports* and *International Aerospace Abstracts*. SCAN, a modified Selective Dissemination of Information (SDI) program, offers 189 subject categories and distributes computer-selected bibliographic listings to users twice monthly. Entries include title, personal and corporate authors, date, and applicable subject descriptors. There are no indexes.

Computer Searches on topics specially requested by the user are available from the NASA Central Data Bank in College Park Md., from any of the seven Regional Dissemination Centers, or by using the NASA/RECON (remote console) service. The remote consoles are installed at a number of locations, including some of the NASA-funded libraries, and are hooked into the central computer and Data Bank in College Park. The libraries and dissemination centers develop local admission conditions of use, but the general policy is to service government officials, NASA contractors, and other personnel on official business.

·SELF-HELPS: The *NASA Thesaurus* (1967). 3 Vols. SP-7030. Available from NTIS or GPO. This is the vocabulary control used for indexing all NASA documents and, therefore, is extensively interdisciplinary, with metallurgical terms widely dispersed. It follows the standard pattern established by the DOD/EJC *Thesaurus*.

USER QUALIFICATIONS: Most of the publications are available for purchase from NTIS or GPO as previously mentioned. Services limited to qualified personnel (NASA personnel, contractors, grantees, and consultants) include retrieval services from the Central Data Bank in Maryland, SCAN, and RECON. However, the Regional Dissemination Centers provide retrieval services on the NASA data base with no user restrictions other than payment of fees, which may vary from a few hundred to several thousand dollars.

C6

NATIONAL BUREAU OF STANDARDS (NBS)
U.S. Department of Commerce
Washington, D.C. 20234

PURPOSE AND FUNCTION: NBS conducts research and provides testing and other services to assure the maximum application of physical and engineering sciences to the advancement of technology in industry and commerce. The Bureau consists of three institutes—Institute of Basic Standards, Institute for Materials Research, and Institute for Applied Technology—in each of which are a number of Divisions. The Institute for Materials Research includes Divisions for Metallurgy and Inorganic Materials, among others. Within the Institute of Basic Standards, NBS operates the Office of Standard Reference Data (see C7). The Office for Information Programs provides information services and disseminates the scientific and technical data generated by the Bureau.

LIBRARY HOLDINGS: Approximately 125,000 volumes.

LIBRARY SERVICES: Open to the public for reference; interlibrary loan. Metallurgical materials are general in nature and deal primarily with heat treatment of metals, refractory materials, electrical metallurgy, metallography, powder metallurgy, and ferrous and nonferrous metallurgy.

PUBLICATIONS: A complete listing of publications can be found in NBS Special Publication 305, *Publications of the National Bureau of Standards, 1966–1967*, together with *Supplement 1* covering the years 1968–1969. Four similar lists are available that cover publications dating back to 1901.

NBS publications should be ordered from the Superintendent of Documents, U.S. Government Printing Office, Washington, D.C. 20402.

Periodicals: NBS Journal of Research. Three sections, available separately: *Physics and Chemistry* (bimonthly); *Mathematical Sciences* (quarterly); and *Engineering and Instrumentation* (quarterly). These sections report NBS research and development in the form of comprehensive scientific papers.

Technical News Bulletin (monthly). Covers the Bureau's research, development, cooperative, and publication activities in summary form.

Nonperiodicals: Handbooks. ·Recommended codes of engineering and industrial practice (including safety codes) developed in cooperation with interested industries, professional organizations, and regulatory bodies.

Special Publications (Formerly *Miscellaneous Publications*). Proceedings of high-level national and international conferences sponsored by NBS; precision measurement and calibration volumes; and other special publications such as administrative pamphlets, wall charts, and bibliographies.

Monographs. Major contributions to the technical literature on various subjects related to the Bureau's scientific and technical activities, often too lengthy for publication in the *Journal of Research.*

NBS Research Highlights. The Bureau's annual report, which includes summaries of the principal NBS research projects in metallurgy.

Products Standards. This series comprises voluntary standards that establish dimensional requirements; technical requirements for products; and methods of testing, grading, and marking.

Technical Notes. A publication medium for communications and reports on data of limited or transitory interest.

An annual bibliographical listing of publications originating in the NBS Metallurgy Division is available from that Division on request.

SPECIAL INFORMATION SERVICES: The Metallurgy Division of the Institute for Materials Research conducts research on the structure and properties of metals, including basic and applied investigations of metallurgical phenomena (precipitation hardening; fatigue and fracture; creep; electrodeposition; stress corrosion; heat treatment of ferrous alloys; fundamental, theoretical, and experimental studies of crystal growth; and imperfections in metal crystals) and the development of new and improved measurement techniques and their application to the study of metals. Within these subject areas, the Division answers inquiries and provides advisory and consulting services, either without charge or under contract to government agencies, the scientific community, and industry. The Metallurgy Division also sponsors two data centers within the National Standard Reference Data System (NSRDS)—the Alloy Data

Center (D4) and the Diffusion in Metals Data Center (D12).

The Inorganic Materials Division has sections on Crystallography and Solid State Physics, provides the same types of services as the Metallurgy Division, and sponsors the Crystal-Data Center (D10).

The Information Section of the Bureau's Office of Engineering Standards Services offers an information service on more than 16,000 engineering and related standards and specifications published by more than 350 U.S. trade, professional, and technical societies. They are catalogued, indexed, and maintained in a technical library. A "keyword-in-context" index has been compiled. The Information Section functions both as a technical library and as a referral service, directing inquirers to the appropriate standards-issuing organizations for copies of published standards.

C7

NATIONAL BUREAU OF STANDARDS
OFFICE OF STANDARD REFERENCE
DATA (OSRD)
U.S. Department of Commerce
Washington D.C. 20234

PURPOSE AND FUNCTION: OSRD administers the National Standard Reference Data System (NSRDS), a nationwide program which provides the U.S. technical community with optimum access to quantitative physical science data, critically evaluated and compiled for convenience. "Standard Reference Data" take the form of numerical values of various properties of materials that are well-defined. Data compilation activities are organized into seven subject categories: nuclear properties, atomic and molecular properties, solid state properties, thermodynamic and transport properties, chemical kinetics, colloid and surface properties, and mechanical properties of materials. OSRD coordinates the activities of about 30 continuing data centers, many of which receive whole or partial financial support from other sponsors. Those of metallurgical interest that report primarily to the Bureau of Standards are:

Alloy Data Center (D4)

Binary Metal and Metalloid Constitution Data Center (D6)

Cryogenic Data Center (D9)

Crystal-Data Center (D10)

Diffusion in Metals Data Center (D12)

Superconductive Materials Data Center (D26).

PUBLICATIONS: *NSRDS News.* A monthly newsletter available on request from OSRD.

Other publications consist of critically evaluated data, critical reviews of the state of quantitative knowledge in specialized areas, computations of useful functions derived from standard reference data, and bibliographies. Form of publication ranges from formal monographs to technical notes and looseleaf data sheets. A complete list of these compilations is available from OSRD.

Three recent publications of particular interest are: *Catalog of Standard Reference Materials*, NBS Special Publication 260, July 1970; *Critical Evaluation of Data in the Physical Sciences—Status Report on the National Standard Reference Data System*, NBS Technical Note 553, June 1970, 73 pp.; and *Annotated Accession List of Data Compilations of the NBS Office of Standard Reference Data*, NBS Technical Note 554, Sept. 1970, 193 pp. The latter is actually a bibliography (with abstracts) of books, journal articles, and reports arranged in the subject categories of OSRD.

OSRD also prepares material for the quarterly *Journal of Physical and Chemical Reference Data* (see American Chemical Society, B5).

SPECIAL INFORMATION SERVICES: An Inquiry Referral Service responds, at present in a limited way, to queries within the scope of the program by providing referrals, references, documentation, or data, as available. The Data System Design and Research Section of OSRD is currently developing mechanized means for data acquisition and handling in NSRDS.

C8

NATIONAL REFERRAL CENTER FOR SCIENCE AND TECHNOLOGY
Library of Congress
Washington, D.C. 20540

PURPOSE AND FUNCTION: The Center "provides a single place to which anyone with an interest in science and technology may turn for advice on where and how to obtain information on specific topics." It does not directly provide technical detail or bibliographic services, but rather refers the requester to those who can. Its scope covers all fields of science and technology, including the engineering sciences, and all kinds of information resources.

PUBLICATIONS: Four directories of information resources, two of which include resources that deal with metallurgical information: *A Directory of Information Resources in the United States: Physical Sciences, Biological Sciences, Engineering,* 1965, 356 pp.; and *A Directory of Information Resources in the United States; Federal Government,* 1967, 419 pp.

INFORMATION SERVICES: Answers inquiries by providing brief descriptions of appropriate information resources. Queries to the Center should include a precise statement of the specific information desired, a statement of the resources already consulted, and a statement of the requester's special qualifications, if any.

USER QUALIFICATIONS: None. Inquiry services are free. The *Directories* may be purchased from the Superintendent of Documents, U.S. Government Printing Office, Washington, D.C. 20402.

C9

NATIONAL RESEARCH COUNCIL OF CANADA (NRC)
Technical Information Service (TIS)
100 Sussex Dr.
Ottawa, Ont., K1A OS3, Canada

PURPOSE AND FUNCTION: The Technical Information Service was established in 1945 to provide small and medium secondary manufacturing industries with free, up-to-date technological information on the properties and processing of materials, efficient operation of manufacturing facilities, new industrial developments, and the results of scientific research. It consists of a central group in Ottawa and 11 field offices across Canada. The Ottawa group is organized into three sections: Technical Enquiry and Answer, Industrial Engineering, and Technological Developments. Metallurgy and metal processing are included in the many and diverse subjects covered.

LIBRARY SERVICES: Utilizes the resources of the National Science Library of Canada (A12).

INFORMATION RETRIEVAL SERVICES: The Technological Developments Section operates a technical awareness or Selective Dissemination of Information (SDI) service which matches user profiles to its magnetic tape files of scientific and technical literature selected by the TIS staff. Checklists of titles and Digests in abstract form are mailed at regular intervals to some 3,000 Canadian companies that have submitted interest profiles. Tech Briefs, on which the abstracts are based, and further follow-up information also are provided.

Titles requested by the 3,000 companies are recompiled into Group Checklists.

SPECIAL SERVICES: The Technical Enquiry and Answer and Industrial Engineering Sections provide consultation services by field officers either on-site or through the Ottawa group.

USER QUALIFICATIONS: NRC-TIS is free to Canadian industry and is not undertaken as a general service.

C10

NATIONAL TECHNICAL INFORMATION SERVICE (NTIS)
U.S. Department of Commerce

Washington, D.C. 20230
(See also JOINT PUBLICATIONS RESEARCH SERVICE, C3)

PURPOSE AND FUNCTION: NTIS supplies the industrial and technical community with unclassified information about government-generated science and technology in defense, space, atomic energy, and other national programs. (Note that Defense Documentation Center, C2, is the government agency responsible for dissemination of DOD "classified" and "limited-distribution" documents, while other agencies, such as AEC and NASA, handle the dissemination of their own classified documents.)

RESOURCES: More than 500,000 unclassified/unlimited government-sponsored scientific and technical reports including those produced by DOD laboratories and contractors. Additions are made at the rate of 40,000 to 50,000 new research reports per year. Metallurgical coverage is indicated in the descriptions of publications and services.

PUBLICATIONS:
Abstracts and Indexes: Government Reports Announcements (GRA), formerly *U.S. Government Research and Development Reports* (USGRDR) (semimonthly). An abstract and announcement bulletin covering reports—including progress reports—of research and development performed under government auspices. Approximately 1,500 reports from all major federal agencies are announced in each issue. Abstracts are arranged according to the COSATI subject fields and groups (see SELF-HELPS below). A guide to the subject field and group structure appears in the introduction of each issue. U.S. Government-sponsored translations and some foreign reports written in English are included. Most documents announced in *GRA* are for sale to the public at a nominal service charge.
Government Reports Index (GRI), formerly *U.S. Government Research and Development Reports Index*

(USGRDI) (semimonthly). A companion publication to *GRA*, it provides subject, personal author, corporate author, contract number, and accession/report number indexes.

Weekly Government Abstracts (WGA) and *Governmental Reports Topical Announcements* (GRTA). Two separate series of announcement bulletins issued by subject, they are similar except that they cover different subject areas, and one is weekly and the other semimonthly. *WGA* is issued in five sections, two of metallurgical interest—*WGA Materials Sciences* and *WGA Environmental Pollution & Control*. *GRTA* announces the remainder of the NTIS input in 30 subject groupings, some containing scattered metallurgical information, such as *GRTA Industrial and Mechanical Engineering* which covers metal processing, bonding, and joining. Gradual conversion of all *GRTA* sections to the *WGA* series is planned.

Fast Announcement Service (FAS). Highlights selected new government reports received by NTIS for public sale. All documents are reviewed for their significance to business and industry; approximately 10 percent are selected. The documents are announced with emphasis, when possible, on commercial application. The announcements are compiled by a subject system of 57 categories. *Fast Announcements* serve as "flash sheets" to focus attention on selected new reports. The frequency of issue depends on new input, selection, and the individual categories chosen by a subscriber. Categories of interest to metallurgists are Bonding/Joining; Cleaning and Finishing; Coatings; Low-Temperature Materials and Processes; Metals, Ferrous and Alloys; Metals, Nonferrous and Alloys; Metalworking and Metal Forming; and Testing, Analysis.

SFCSI (Special Foreign Currency Science Information Program) *List of Translations in Process* (annual). A listing, by country, of translations that are being prepared by foreign nationals through the use of U.S.-owned foreign currencies. After the translations are completed, they are announced in *GRA*. The program is administered by the National Science Foundation.

INFORMATION RETRIEVAL SERVICES: Computer magnetic tapes of the data base composing *GRA* are available on a current awareness basis, updated semimonthly. Tapes of *U.S. Government Research and Development Reports* for the year 1970 also are available.

NTISearch is a retrospective searching service based on all of these tapes. Questions are reviewed by an expert information analyst for computer processing. The fee is moderate.

A current awareness or alerting service is provided by subscribing to 1 or more of the 321 NTIS subcategories. Searching services based on the *GRA* tapes are also available from a number of information dissemination centers as listed in directory section D.

Selected Categories in Microfiche (SCIM). A standing-order or current awareness service that provides the customer with microfiche copies of NTIS reports. The service may be ordered by category or subcategory, agency collection (DOD, NASA, or AEC reports, for example), or subcategories within an agency collection.

SELF-HELPS: The *COSATI Subject Category List*, AD 612 200, is a two-level arrangement consisting of 22 major subject fields with a further subdivision into 188 subject groups. Subject field No. 11, "Materials," contains 12 groups, 4 of which are of metallurgical interest: Metallurgy and Metallography; Ceramics, Refractories, and Glasses; Coatings, Colorants, and Finishes; and Composite Materials. Other peripheral topics are scattered throughout some of the other groups; for example, field No. 20, "Physics," contains groups on Crystallography, Solid State Physics, and Thermodynamics.

C11

SCIENCE INFORMATION EXCHANGE
(SIE)
Smithsonian Institution
300 Madison National Bank Bldg.
1730 M St., N.W.
Washington, D.C. 20036

PURPOSE AND FUNCTION: A national registry of research in progress. SIE is partially supported by the National Science Foundation, with the cooperation and/or participation of more than 1,000 research organizations, including federal agencies, private foundations, universities, state and city governments, industry, and foreign sources. SIE complements the services of technical libraries and documentation centers by providing information about research in progress from the time a project is proposed or started to the time that results are published.

RESOURCES: Approximately 85,000 to 100,000 projects per year, in all fields of research, are being registered at the Exchange. The number varies according to increased or decreased levels of support for research. Information is registered on a 1-page Notice of Research Project (NRP). This is the basic document of the Exchange and includes the name of the granting agency, names and addresses of principal and associate investigators, location of the work, title, a 200-word summary of technical detail, and the level of effort. The information is indexed, cross-referenced, and coded into a computer.

METALLURGICAL COVERAGE: The data bank currently contains between 2,000 and 3,000 ongoing research summaries dealing with metals and alloys. Typical examples of subjects that have been searched and the number of NRP's retrieved are:

Ferrous metals, approximately 450
Refractory and heat resistant alloys, 300
Stress corrosion, 75
Welding, 150
Powder metallurgy, 100
Unconventional machining, 40
Alloys of cobalt, 45.

PUBLICATIONS: *Science Research in Progress—1972*. A 12-volume series describing more than 125,000 current research projects. Indexed by subject, principal investigator, research organization, and source of supporting funds. One of the volumes, *Materials*, describes approximately 4,000 projects. Published by Academic Media, 32 Lincoln Ave., Orange, N.J. 07050.

INFORMATION RETRIEVAL SERVICES: To use SIE, prepare a "Request for Services" form, available from SIE, stating the specific research or problem on which information is desired. The Exchange will forward promptly pertinent NRP's. Types of output services are: Subject Searches—comparable to a bibliographic compilation; Selective Dissemination or Periodic Mailing—research notices sent quarterly, based on a profile of the requester's interest; Administrative Content Searches—retrieval of all records related to a given country, state, institution, or department; Investigator Search—response to a request for research associated with a given investigator's name; and Standard Computer Tabulations and Listings—project listings based upon given selection criteria and in various sequences. Data for each project may include the supporting organization, principal investigator, agency contract and/or accession numbers, performing organization, location, school, department, project title, and funding.

USER QUALIFICATIONS: None; price schedule (relatively low) available on request.

C12

SUPERINTENDENT OF DOCUMENTS
U.S. Government Printing Office
(GPO)
Washington, D.C. 20402

PURPOSE AND FUNCTION: The agency authorized by law to sell copies of government publications. There is no general price list of public documents, but numerous lists have been prepared on special subjects. Any of these lists will be furnished free, on application, by subject or subjects of the information desired. Free government documents generally are available from the issuing agencies.

PUBLICATIONS: *Monthly Catalog, U.S. Government Publications.* A current bibliography of publications issued by all branches of the government, including both congressional and departmental publications. Entries are arranged alphabetically by issuing agency and include title, author, date, pagination, price, availability, and other information. Symbols indicate whether a publication is for sale by the Superintendent of Documents, the National Technical Information Service, or distributed by the issuing office. Monthly indexes appear in each issue; annual index in the December issue. The February issue contains the *Directory of United States Government Periodicals and Subscriptions.*

C13

U.S. ATOMIC ENERGY COMMISSION (AEC)
DIVISION OF REACTOR DEVELOPMENT AND TECHNOLOGY
Washington, D.C. 20545

Sponsors the Liquid Metals Information Center (D18), an AEC specialized information and data center.

C14

U.S. ATOMIC ENERGY COMMISSION (AEC)
DIVISION OF RESEARCH
Washington, D.C. 20545

Sponsors two specialized information and data centers: Research Materials Information Center (see D25) and Thermodynamic Properties of Metals and Alloys (see D29).

Publishes *Research Contracts in the Physical Sciences,* an annual listing of contract research projects supported by the AEC Headquarters Division of Research. Arranged alphabetically within several broad subject areas, it contains name and address of contractor, name(s) of

principal investigator(s), a short descriptive title of the research, and the level of support during the most recent funding period. Also included are a list of major AEC research centers with their level of support for the current fiscal year; a summary of off-site contracts by type of organization and subject areas; a summary of new proposals received and actions taken; and a summary, by state, of numbers of contracts awarded and amounts of funding.

C15

U.S. ATOMIC ENERGY COMMISSION (AEC)
DIVISION OF TECHNICAL INFORMATION (DTI)
P.O. Box 62
Oak Ridge, Tenn. 37830

PURPOSE AND FUNCTION: To alert working scientists and engineers to the international nuclear science literature and to abstract, index, and distribute such material; to publish scientific books, monographs, and technical progress review journals; to provide educational literature on nuclear energy to a wide audience, generally on the high-school level; to offer nuclear science reference services by establishing specialized information centers; to sponsor scientific conferences for the exchange of ideas among scientists and engineers; to channel information to potential industrial users of AEC-generated technical innovations; to promote knowledge of the peaceful uses of nuclear energy through demonstrations to high-school and public audiences; and to foster international cooperation in the development of nuclear applications by participating in information activities with foreign and international organizations. Serves as the distribution center for AEC technical report literature.

METALLURGICAL COVERAGE: The wide range of AEC technical interests is reflected in the 10 subject sections of *Nuclear Science Abstracts,* 1 of which is devoted to "Metals, Ceramics and Other Materials." This section contains nearly 8 percent of the total output—more than 4,000 abstracts annually. It includes corrosion and erosion, preparation and fabrication, properties evaluations, structural and phase studies, and radiation effects, insofar as these topics are associated with nuclear technology. The "Chemistry" section includes analytical techniques, and the "Engineering" section includes materials testing.

LIBRARY HOLDINGS: The collection contains 4,000 books and bound journals, 1,250 periodical subscriptions, and 400,000 reports.

LIBRARY SERVICES: Interlibrary loan. The AEC Library is not open to the public, but employees of federal agencies and their contractors, professors, graduate students, and industrial employees can make prior arrangements to use the Library. Copies of AEC reports are available on microfiche or microcard. Sets currently are being provided to more than 100 libraries in the United States and throughout the world. A list of these libraries is published in the booklet, *Science Information Available from the Atomic Energy Commission.*

PUBLICATIONS:

Books: The catalog, *Technical Books and Monographs*, lists a large number of titles under 12 subject headings. The section on metallurgy contains 40 titles; others of metallurgical interest are listed in the sections on chemistry, engineering and instrumentation, reactors, and vacuum technology. A number of monographs have been prepared for AEC by technical societies; for example, the American Society for Metals has sponsored 13 monographs on metallurgical topics. Many of the books are available from commercial publishers.

Abstracts and Bibliographies: Nuclear Science Abstracts (NSA) (semimonthly, 1948–). Available from the Superintendent of Documents, U.S. Government Printing Office, Washington, D.C. 20402. Approximately 50,000 abstracts per year in 10 subject categories (see METALLURGICAL COVERAGE above) from 2,000 journals, plus government and industry reports, books, conference proceedings, patents, and dissertations. Each issue contains corporate author, personal author, subject, and report number indexes, cumulated quarterly, annually, and approximately every five years.

Bibliographies of Atomic Energy Literature are issued at frequent intervals and contain listings of domestic and foreign bibliographies recently completed or in progress. Available from AEC/DTI at no charge.

Other: The following helpful brochures are available from AEC/DTI.

Science Information Available from the Atomic Energy Commission. A general listing of publications, reference tools, data centers, educational materials, and other services.

Technical Books and Monographs. An indexed catalog of publications sponsored by AEC.

Guide to Nuclear Science Abstracts. A description of the journal and instructions on how to use it.

Subject Scope of Nuclear Science Abstracts. The best source of information on metallurgical coverage of AEC services; Part 2, a list of the elements covered and the extent of interest in each, is particularly useful.

Directory of U.S. Atomic Energy Commission Librarians and Information Specialists. Includes libraries of contractors as well as AEC divisions.

Directory of USAEC Specialized Information and Data Centers. A list of about 25 centers. Examples of three centers with metallurgical interests are:

Liquid Metals Information Center (D18)

Research Materials Information Center (D25)

Thermodynamic Properties of Metals and Alloys (D29).

SPECIAL INFORMATION RETRIEVAL SERVICES: The indexes to *Nuclear Science Abstracts* (NSA) are generated by computer, and the data base tapes are used by a limited number of subscribers on an experimental basis. AEC/DTI also offers a subject search by computer going back to 1962 and covering 500,000 items. The product is a computer printout of NSA abstract numbers and relevant subject indexing. The service is offered to the general public at a moderate fee; the price to government agencies, contractors, and grantees is somewhat lower. AEC is also experimenting with the NASA/RECON system at five locations which feed into the central computer at Oak Ridge. (See Lawrence Radiation Laboratory, D17.)

SELF-HELPS: Two publications are useful in organizing personal or corporate files. *Subject Headings Used by the USAEC Division of Technical Information*, TIP-5001, is the authority for subject headings used in the published journal, *NSA*. This indexing authority contains 19,500 subject headings, 8,500 *see* cross-references, and approximately 10,000 *see also* cross-references. *INIS Thesaurus*, IAEA-INIS-13 (available from the Division of Scientific and Technical Information, International Atomic Energy Agency, P.O. Box 590, A-1011, Vienna, Austria), is the authority for assigned keyterms in *NSA* and also for SDI and retrospective searching by *NSA* tape users. The *Thesaurus* contains approximately 16,000 keyterms.

USER QUALIFICATIONS: Qualifications will vary depending upon product or service; but, in general, there are few restrictions.

C16

U.S. BUREAU OF MINES
Department of the Interior
18th and C Sts., N.W.
Washington, D.C. 20240

PURPOSE AND FUNCTION: Established in the public interest to conduct inquiries and scientific and technologic investigations concerning mining and the preparation, treatment, and utilization of mineral substances; to promote health and safety in the mineral industries; to conserve material resources and prevent their waste; to further economic development; to increase efficiency in the mining, metallurgical, quarrying, and other mineral industries; and to inquire into the economic conditions affecting these industries. The Bureau maintains a number of research centers dealing with various subject interests, including the following for metallurgy:

Albany Metallurgy Research Center, P.O. Box 70, Albany, Ore. 97321. Areas of interest: production of titanium, zirconium, and hafnium; melting, casting, and forming of high-purity refractory metals; mineral dressing; pyrometallurgy; hydrometallurgy; and recovery of metals from industrial wastes and scrap metal products.

Boulder City Metallurgy Research Laboratory, 500 Date St., Boulder City, Nev. 89005. Areas of interest: electrowinning and electrorefining of refractory and reactive metals, preparation of light metals by metallic reduction techniques, and recovery of metals from scrap and residues.

College Park Metallurgy Research Center, College Park, Md. 20740. Areas of interest: corrosion, recycling waste and scrap metals, flotation of base metal sulfides, analysis and characterization of less common metals and minerals, and magnetic and electrostatic separation of metals and minerals.

Reno Metallurgy Research Center, 1605 Evans Ave., Reno, Nev. 89505. Areas of interest: electrometallurgical processing of rare-earth metals and alloys; hydrometallurgical and solvent extraction processes for recovering copper, silver, gold, mercury, and other metals; recovery of valuable metals from scrap; and development of advanced analytical techniques for identifying precious metals.

Rolla Metallurgy Research Center, P.O. Box 280, Rolla, Mo. 65401. Areas of interest: mineral dressing, extractive metallurgy, and physical metallurgy.

Salt Lake City Metallurgy Research Center, 1600 E. First South St., Salt Lake City, Utah 84112. Areas of interest: recovery of copper, molybdenum, uranium, magnesium, silver, and heavy metals.

Tuscaloosa Metallurgy Research Laboratory, P.O. Box L, University, Ala. 35486. Areas of interest: beneficiation and utilization of nonmetallic minerals and materials including mica, phosphate, fluorspar, clay, and waste glass from municipal refuse.

Twin Cities Metallurgy Research Center, P.O. Box 1660, Twin Cities Airport, Minn. 55111. Areas of interest: extractive metallurgy with emphasis on iron, copper, and physical chemistry.

LIBRARY SERVICES: All of the research facilities maintain small libraries of pertinent publications that are open to the public for reference. In addition, representative collections of Bureau of Mines publications are housed at nearly 700 depository libraries in the United States. A list of these libraries, indicating types of publications held, is given in *List of Bureau of Mines Publications and Articles*.

PUBLICATIONS: Three catalogs of Bureau of Mines publications are available. *List of Bureau of Mines Publications and Articles:* January 1, 1965 to Dec. 31, 1969, with subject and author index; a list for January 1960 through December 1964; and a 50-year list covering July 1910 to January 1960. A monthly list of *New Publications—Bureau of Mines* is also available from the Bureau and indicates where publications may be obtained.

Types of publications are as follows:

Bulletins. Describe major Bureau investigations or studies that are considered to have permanent value.

Minerals Yearbook. Annual statistical publication which reviews the mineral industry in the United States and foreign countries.

Reports of Investigations. Describe the principal features and results of minor investigations or phases of major investigations, thus keeping the public informed on the progress of original research.

Information Circulars. Easily understood digests designed primarily for compilations, reviews, abstracts, and discussion of virtually all activities and developments in the mineral industries.

Technical Progress Reports. A series, initiated in 1968, reporting new or improved systems and techniques in mining and metallurgy developed by the Bureau.

Mineral Industry Surveys. Cover a wide variety of timely statistical and economic reports.

Foreign Mineral Reports. Keep the domestic producers and consumers abreast of developments and markets abroad.

Special Publications. Include comprehensive lists of Bureau publications, articles by Bureau authors, and any publication of the Bureau of Mines not included in its regular series.

Schedules. Describe the procedures and methods followed by the Bureau in testing materials and equipment to determine their permissibility for use by the mineral industries.

Handbooks. Special manuals to promote safety and efficiency in the mineral industries.

Miscellaneous Publications. Include any Bureau of Mines publication that is not part of a Bureau series.

Availability: Some Bureau publications, including the catalogs, *Bulletins,* and the *Minerals Yearbook,* are sales publications; other series contain both free and sales publications. Sales publications may be obtained from the Superintendent of

Documents, U.S. Government Printing Office, Washington, D.C. 20402. Free publications may be obtained from the Publications Distribution Branch, Bureau of Mines, 4800 Forbes Ave., Pittsburgh, Pa. 15213. Most new reports, in either paperback or on microfiche, are available from the National Technical Information Service, U.S. Department of Commerce, Springfield, Va. 22151.

Additions to mailing lists for new metallurgical publications and the monthly list of *New Publications* should be addressed to the Chief, Division of Technical Reports, Bureau of Mines, U.S. Department of the Interior, Washington, D.C. 20240.

C17

U.S. PATENT OFFICE

U.S. Department of Commerce
Washington, D.C. 20231

PUBLICATIONS: *Manual of Classification, U.S. Patent Office.* Available from the Superintendent of Documents, U.S. Government Printing Office, Washington, D.C. 20402. Copies of patents by subscription to 1 or more of 346 main classes or 60,000 subclasses in the *Manual of Classification.*

Official Gazette of the United States Patent Office (weekly, 1872–). Approximately 50,000 abstracts per year arranged in three broad groups: General and Mechanical, Electrical, and Chemical. An alphabetical name index of patentees and classified subject indexes in each issue, Also available from the Superintendent of Documents.

D

INFORMATION CENTERS

D1

AEROSPACE MATERIALS INFORMATION CENTER (AMIC)
Wright-Patterson AFB, Ohio 45433

The description of this agency was deleted at press time because of a change in policy. While AMIC provided useful services for many years to government agencies, universities, and industry engaged in the defense effort, its services now are limited to personnel of the Air Force Materials Laboratory.

D2

AEROSPACE RESEARCH APPLICATIONS CENTER (ARAC)
Indiana University Foundation
Indiana Memorial Union
Bloomington, Ind. 47401

SPONSORS: Indiana University, in cooperation with the National Aeronautics and Space Administration. Serves as a NASA Regional Dissemination Center.

PURPOSE AND FUNCTION: A nonprofit information center providing a variety of computerized literature searching services. Its goals are to promote the rapid application of new technology and to reduce the duplication of expensive research.

RESOURCES: Searchable magnetic tape data files from government and private sources as follows: NASA

Scientific and Technical Aerospace Reports (C5), *International Aerospace Abstracts* (B10), *Government Reports Announcements* (C10), *Reliability Abstracts and Technical Reviews, Engineering Index* COMPENDEX (F4), and *CA* Condensates (F3).

INFORMATION RETRIEVAL SERVICES: There are six SDI or current awareness services, three being of technical metallurgical interest and three of management and marketing interest. The former are Standard Interest Profiles, Custom Interest Profiles, and Industrial Applications Service. Standard Interest Profiles provide continuous monitoring of the literature on a specific subject. Output is in the form of packages of abstracts or citations that are mailed once or twice a month. Under the heading of "Materials," the ARAC catalog lists 17 profiles of metallurgical interest, ranging from crystal growth to welding and cutting. Searches of the government data bases are subscribed to separately from those of COMPENDEX and *CA* Condensates; in some instances the same topic is available from both sources. Custom Interest Profiles are similar to Standard Interest Profiles, except that they are personalized. Cost is dependent upon the number and scope of subjects covered, literature sources searched, and degree of difficulty encountered.

D3

AIR FORCE MACHINABILITY DATA CENTER (AFMDC)
[Now known as MACHINABILITY DATA CENTER (MDC)]
Metcut Research Associates, Inc.
3980 Rosslyn Dr.
Cincinnati, Ohio 45209

SPONSOR: Department of Defense, Defense Supply Agency.

PURPOSE AND FUNCTION: MDC collects, evaluates, stores, and disseminates material removal information, including specific and detailed machining data, for the benefit of industry and government. Strong emphasis is given to engineering evaluation for the purpose of developing optimized material removal parameters. Data are being processed for all types of materials and for all material removal operations, including conventional machining and alternate removal processes.

RESOURCES: A data file of more than 32,000 selected documents from journals, periodicals, industrial literature, government reports, and technical correspondence pertaining to all phases of material removal technology. The monthly accession rate is about 300 documents. Information from these resources—consisting of specific materials, their chemical, physical, and mechanical properties, and the specific material removal operations being used—is indexed and coded for a computerized retrieval service.

LIBRARY FACILITIES: The above resources are available for reference and for direct and interlibrary loan. Photocopy service is also provided.

PUBLICATIONS:
Machining Data Handbook, 2nd Edition, 1972.
A number of other data publications are available which cover machining of titanium, beryllium, and high-strength steels, as well as numerical control and computer applications to machining.

INFORMATION RETRIEVAL SERVICES: Qualified users can register their interests in detail on a special form. They then receive information products which include machining data pamphlets, tables on materials of current interest, state-of-the-art reports, technical announcements, and other appropriate items.

SPECIAL SERVICES: Consultation on machinability problems; translated abstracts of certain French and German articles.

USER QUALIFICATIONS: Services are available to the aerospace industry, all DOD agencies and their contractors, other government agencies, technical institutions, and nonmilitary industries in a position to assist the defense effort.

D4

ALLOY DATA CENTER (ADC)
National Bureau of Standards
Institute for Materials Research
Washington, D.C. 20234

SPONSOR: NBS Metallurgy Division, Alloy Physics Section.

PURPOSE AND FUNCTION: ADC's purposes are to stimulate communication and exchange of information among existing data centers in the field of physical properties of metals and alloys and to collect and evaluate data in those areas where special competence exists in the NBS Alloy Physics Section.

LIBRARY HOLDINGS: Collection of 10,000 references to research papers.

LIBRARY SERVICES: Open to the public for reference; loans; self-service photocopy service.

PUBLICATIONS:
Permuted Materials Index, NBS Special Publication 324 (available from the U.S. Government Printing Office). All records are arranged alphabetically by chemical symbol. Records referring to alloys or compounds are listed under each of the constituent elements.
Author Index, Office of Standard Reference Data Bibliography Series, OSRD-B-70-2. All papers are listed

alphabetically by first author in three groupings: (a) nuclear magnetic resonance papers, (b) soft x-ray emission spectra in metals and alloys, and (c) papers of general interest to the Alloy Physics Section. Handbooks on the first two topics are in preparation.

INFORMATION RETRIEVAL SERVICES: The *Permuted Materials Index* and the *Author Index* are recorded on magnetic tape, and computer searches are offered at a fee based solely on reimbursement of computer-run cost. Special indexes and bibliographies can be prepared on the same basis. General information requests that can be handled without computer runs also are answered.

D5

ARMY MATERIALS AND MECHANICS RESEARCH CENTER (AMMRC)
Technical Information Branch and Technical Library
Watertown, Mass. 02172

SPONSOR: Department of Defense, U.S. Army.

PURPOSE AND FUNCTION: To provide direct service to scientists and engineers of the Center by taking care of their information needs. Areas of interest are metals—their compounds, mixtures, and alloys; metallography; powder metallurgy; refractories; composites; industrial processes; test and evaluation; specifications and standards; ceramics and glasses; solid state physics; crystallography; structural engineering; solid mechanics; and materials related to armor, armament, ballistics, and aircraft as used by the Army.

LIBRARY HOLDINGS: 30,000 books, 40,000 documents, and 800 serial publications.

LIBRARY SERVICES: A full range of library services including retrieval and dissemination of government, private, foreign, and domestic books; serials and periodicals; documents; abstracts and indexes; and vertical file material. Interlibrary loan.

PUBLICATIONS: Technical Reports, Technical Memoranda, Technical Notes, Monographs, and Special Publications.

INFORMATION SERVICES: Technical reference, literature searches, and indexing. Also operates the Nondestructive Testing Information Analysis Center (D22).

USER QUALIFICATIONS: Facilities are limited to DOD agencies and laboratories and other authorized DOD contractors and subcontractors.

D6

BINARY METAL AND METALLOID CONSTITUTION DATA CENTER (BMMCDC)
IIT Research Institute
10 W. 35th St.
Chicago, Ill. 60616

SPONSORS: National Bureau of Standards, Office of Standard Reference Data; Atomic Energy Commission; and Aerospace Research Laboratories.

PURPOSE AND FUNCTION: BMMCDC's mission is to prepare reviews of published constitution data similar to the work initiated by M. Hansen in *Constitution of Binary Alloys*, McGraw-Hill, 1958. Scope primarily is binary combinations of metallic elements, excluding binaries with halogens and those which are not metal-related. Properties covered are solidus, liquidus, invariant reactions, solubility, crystal structure, and lattice spacings. Parameters covered are temperature and composition at pressures near the ambient.

INFORMATION RETRIEVAL SERVICES: Although the primary purpose of this Center is to search and review the literature and not to serve as an information center, requests for references to a limited number of specific binary systems can be processed. For more extensive searching, the Center's files are available to individual searchers by appointment, or a research program can be initiated at the IIT Research Institute for a comprehensive search and evaluation by the staff.

USER QUALIFICATIONS: U.S. Government agencies and their contractors, research and educational institutions, and industry.

D7

COBALT INFORMATION CENTER
c/o Battelle Memorial Institute
505 King Ave.
Columbus, Ohio 43201

SPONSOR: U.S. affiliate of Centre d'Information du Cobalt, S.A., Brussels, Belgium.

PURPOSE AND FUNCTION: A nonprofit organization maintained to provide technical assistance to cobalt users, disseminate technical information, encourage and aid research, and develop new uses for cobalt. The Center covers physical, mechanical, and chemical metallurgy and the metallurgical engineering aspects of cobalt.

LIBRARY SERVICES: Resources of the Battelle library, together with an extensive special collection, are available to registered visitors for reference.

PUBLICATIONS:
Cobalt (quarterly). Publishes reports on research conducted by the Center, plus technical articles. Includes a section, *Review of Technical Literature*, which publishes approximately 400 abstracts per year, arranged in 10 subject categories.
Cobalt Monograph, 1960. Currently being updated. The first supplement, on cobalt-base superalloys, appeared in 1970; work is in process on a second supplement on magnetic materials and a third supplement on high-strength steels.
Other publications include reprints, special reports, occasional bibliographies, and data sheets.

SPECIAL INFORMATION SERVICES: Answers to technical inquiries; literature searches; consultation either at Battelle or by visiting experts; and lectures and films provided to technical and education groups.

USER QUALIFICATIONS: Services are free to those having a genuine interest in the use of cobalt.

D8

COPPER DATA CENTER (CDC)
Battelle Memorial Institute
505 King Ave.
Columbus, Ohio 43201

SPONSOR: Copper Development Association, Inc. (B31).

PURPOSE AND FUNCTION: Main objectives of the Center are to provide engineers, who select and apply materials, with technical data on the properties and applications of copper, brass, and bronze and to provide CDA member companies with immediate access to world information on the technology of copper and copper alloys. The Center's scope covers copper technology from refining of the metal to the end-use performance of parts, components, and systems made from its mill products. Materials included are copper and copper alloys, iron and steel with copper as an alloying element, copper chemicals, materials that compete with copper applications, and materials and processes used in insulated wire and cable.

RESOURCES AND PUBLICATIONS: A computer-based information retrieval system is built on the following publications: *Extracts of Documents on Copper Technology*, a *Coordinate Index* to the *Extracts* (irregular, 1965–), plus a *Thesaurus of Terms on Copper Technology*.

INFORMATION RETRIEVAL SERVICES: The Center responds to technical inquiries, by mail or phone, and prepares data sheets and technical reports. A computer-coded field-of-interest register is maintained for distribution of data sheets, technical reports, and other special publications.

A current awareness (SDI) service is provided, whereby a batch computer run matches the keywords of user interests to those of new documents being entered in the system. The Center also has the capability of conducting coordinate searches of the system using remote terminals connected on-line to a time-sharing computer located at Battelle Columbus Laboratories. This interactive system is part of BASIS-70 (Battelle Automated Search Information System).

USER QUALIFICATIONS: Technical inquiries are answered without charge. Distribution of the three-part information package (*Extracts*, *Index*, and *Thesaurus*) and the SDI service are limited to CDA member companies.

D9

CRYOGENIC DATA CENTER (CDC)
National Bureau of Standards
Boulder Laboratories
Boulder, Colo. 80302

SPONSOR: National Bureau of Standards, Office of Standard Reference Data.

PURPOSE AND FUNCTION: Designed to serve as the major source of bibliographic information and data on the properties of materials at cryotemperatures. CDC is engaged in the critical evaluation and compilation of data on thermodynamic, transport, and other thermophysical properties of the principal fluids used at low temperatures. The scope also includes the following properties of metallic elements, selected alloys, and element dielectrics: electrical resistivity, dielectric constant, thermal conductivity, thermal expansion, specific heat, and enthalpy. Ultimately, it is expected that data will be compiled for the mechanical properties of structural materials.

LIBRARY SERVICES: More than 64,000 accessions of cryogenic literature have been entered in the system. Approximately 35,000 of these, on properties of both fluid and solid materials, have been processed for machine searching. Visitors are invited to use the Center's library, world literature file, catalog and abstract files, and microfilm facilities.

PUBLICATIONS:
NSRDS (National Standard Reference Data System) *Monographs*.
NBS (National Bureau of Standards) *Technical Notes*.
Announcements of Cryogenic Lab-

oratory Publications and Reports. Available to anyone wishing to be placed on the mailing list.
Superconducting Devices and Materials Literature Survey (quarterly). Published in cooperation with the Office of Naval Research. The March 1970 issue (a typical example) included 145 references indexed under 39 subject headings. Available from the Center at a subscription fee.
Reports on properties of materials; occasional bibliographies on various materials and properties; tables and charts of thermophysical property data. Approximately 600 items are available from either the National Technical Information Service (C10) or the U.S. Government Printing Office (C12).

INFORMATION SERVICES: Current Awareness Service (weekly). Lists of new literature of cryogenic interest, available from the Center at a subscription fee.
Custom Bibliographies. Detailed and extensive bibliographies prepared with computer facilities.
Consultation service is also provided.

D10

CRYSTAL-DATA CENTER (CDC)
National Bureau of Standards
Institute for Materials Research
Washington, D.C. 20234

SPONSOR: National Bureau of Standards, Inorganic Materials Division.

PURPOSE AND FUNCTION: CDC's major mission is the revision of *Crystal Data Determinative Tables*. The data consist of axial lengths and angles of the unit cell, space groups, number of molecules or formula weights per cell, both the measured densities and those calculated from x-ray data, habits, cleavages, twinnings, colors, refractive indices, and melting points.

PUBLICATIONS: *Crystal Data*, 2nd Edition, 1963. Published by the American Crystallographic Association and available from the Polycrystal Book Service, P.O. Box 11567, Pittsburgh, **Pa.** 15238. The 3rd edition, published in 1971, contains 30,000 entries (13,000 in the 2nd edition) in bibliographic form with references to the original literature.

D11

DEFENSE METALS AND CERAMICS INFORMATION CENTER (MCIC)
Battelle Memorial Institute
Columbus Laboratories
505 King Ave.
Columbus, Ohio 43201
(Until 1972, existed as two separate centers—DEFENSE METALS INFORMATION CENTER and DEFENSE CERAMIC INFORMATION CENTER)

SPONSOR: Department of Defense, Defense Supply Agency.

PURPOSE AND FUNCTION: MCIC's objective is to provide a comprehensive, current resource of technical information on the advanced metals and ceramics. The advanced metals consist of aluminum and magnesium, beryllium, titanium, refractory metals, high-strength steels, and superalloys. For these metals and their alloys, MCIC provides current information on production, primary fabrication, secondary fabrication, powder metallurgy, composites, joining, mechanical and physical properties, physical metallurgy, surface treatment, environmental effects, and quality control. Coverage of ceramics includes borides, carbides, carbon (graphite), nitrides, oxides, sulfides, and silicides, as well as selected intermetallic compounds, metalloid elements and glasses, coatings, fibers, composites, and foams. The subject field encompasses applications; property and performance data; and processing, fabricating, and testing methods. Ceramics coverage is slanted primarily toward structural and thermal-protective applications in aerospace and other military systems.

RESOURCES: The Center's collection consists primarily of journals, periodicals, government reports, industrial literature, technical and scientific data, and technical correspondence, including classified material. The holdings are primarily in the form of evaluated extracts from the original literature and are thoroughly indexed. The technical files are stored both on computer, for time-sharing retrieval, and on a multiple-indexed manual card system. The resources generally are not open to the public.

PUBLICATIONS:
Reviews of Recent Developments (quarterly). Presents brief summaries of information on 16 separate areas of metals and processes which have become available in the preceding period.
State-of-the-Art Reports. Comprehensive studies of the current status of selected metals, alloys, or processes.
Technical Memoranda. Preliminary summaries of recent developments in specific aspects of metals, alloys, or processes.
Technical Notes. Informal collections of timely information on topics of special interest.
Handbooks. Condensed summaries of current information on the properties and processes of selected metals.
Ceramic Awareness Bulletin (bimonthly). Provides informative digests of selected accessions, briefings on significant technical advances, and listings of new or continuing government R&D programs.
Ceramic R&D Programs (annual). Provides descriptive listings of current government ceramics projects.
An *MCIC Newsletter* is distributed free of charge, but all other publications are distributed by the National Technical Information Service (C10); prices are on a cost-recovery basis.
Engineering Properties of Ceramics (AFML-TR-66-52). A looseleaf data book, updated with new sections as required, is produced by the Center for distribution by the American Ceramic Society (B4).

SPECIAL INFORMATION SERVICES: Prepares critical reviews and state-of-the-art studies; answers technical inquiries; provides technical advisory service and literature searches; and conducts surveys and limited research investigations. Fees are based on the amount of engineering time involved in responding to inquiries.

D12

DIFFUSION IN METALS DATA CENTER (DMDC)
Institute for Materials Research
National Bureau of Standards
Washington, D.C. 20234

PURPOSE AND FUNCTION: To collect and evaluate data on diffusion in metals and alloys.

LIBRARY AND LIBRARY SERVICES: The collection, which includes copies of approximately 8,000 papers from the scientific literature that report data on diffusion in metals, is open to the public for reference. Answers specific information requests that can be handled by available staff.

PUBLICATIONS: A series of monographs is in preparation. Each monograph will summarize and present critical analyses of diffusion data on a particular set of alloy systems. It is planned to cover diffusion in all pure metals and binary alloys.

SPECIAL INFORMATION SERVICES: Translations of Russian diffusion papers prepared for the Center can be purchased by the public through the National Technical Information Service. (See *SFCSI List of Translations in Process*, NTIS, C10.)

D13

DOW CURRENT AWARENESS SERVICE (DCAS)
Building 1707
The Dow Chemical Co.
Midland, Mich. 48640

PURPOSE AND FUNCTION: A computerized information retrieval service originally developed for use by Dow Chemical Co. personnel, but now offered to the general public on a fee basis.

RESOURCES: Searchable magnetic tape data files including *CA* Condensates and *Chemical Titles* (F3)

and *Engineering Index* COMPENDEX (F4).

INFORMATION RETRIEVAL SERVICES: Current awareness service on an individual interest profile basis only. Computer printouts of pertinent literature references are delivered every two weeks.

USER QUALIFICATIONS: None. Fees are negotiated individually. *Profile Preparation Manual* available on request.

D14

ELECTRONIC PROPERTIES INFORMATION CENTER (EPIC)
Hughes Aircraft Co.
Centinela and Teale Sts.
Culver City, Calif. 90230

SPONSOR: Department of Defense, Defense Supply Agency.

PURPOSE AND FUNCTION: EPIC collects, indexes, and abstracts the world's literature on the electrical, magnetic, and optical properties of materials of value to the defense community and evaluates, compiles, and publishes the experimental data from this literature. Ten major categories of materials are covered: semiconductors, insulators, metals, superconductors, ferromagnetics, ferro-electrics, ferrites, electroluminescents, thermionic emitters, and optical materials.

RESOURCES: More than 100 technical journals, several abstract journals, and the unclassified report literature are searched regularly for pertinent information. Some 48,000 documents have been acquired by the Center. They are indexed in detail for a computer-based retrieval system.

PUBLICATIONS:
EPIC Bulletin (quarterly). Announces new publications and current activities.
Handbook of Electronic Materials (Plenum Press), three volumes.
Electronic Properties of Materials —A Guide to the Literature (Plenum Press), three volumes.
EPIC Data Sheets. State-of-the-art reports and bibliographies; approximately 100 titles.

EPIC Interim Reports. Informal compilations of data and bibliographic information prepared in response to technical inquiries and considered to be of general interest; approximately 40 titles.

INFORMATION RETRIEVAL SERVICES: Computer searches provide bibliographies and comprehensive, critically evaluated compilations of the electronic, optical, and magnetic properties of materials.

USER QUALIFICATIONS: None. Publications can be ordered from EPIC, National Technical Information Service (C10), or Plenum Press. Technical inquiries are priced at an hourly labor rate. A price list for publications and computer searches is available on request.

D15

IIT RESEARCH INSTITUTE (IITRI)
Information Sciences Section
10 W. 35th St.
Chicago, Ill. 60616

PURPOSE AND FUNCTION: The Information Sciences Section of IITRI is engaged in basic and applied research in the disciplines associated with the acquisition, storage, processing, and dissemination of information. The Computer Search Center was established within the Information Sciences Section in 1969. It is a source for computer SDI and retrospective searches of machine-readable data bases.

RESOURCES: Searchable magnetic tape data files include, among others, *CA* Condensates (F3), *Engineering Index* COMPENDEX (F4), and Institute for Scientific Information's ASCA (E10). The John Crerar Library (A7) is located on the IIT (Illinois Institute of Technology) campus, and its resources are available for backup in the form of published abstract services and original documents.

INFORMATION RETRIEVAL SERVICES: Both retrospective and current awareness (SDI) searches. Results are disseminated in the form of abstract cards, punched paper tapes or IBM cards, drilled Termatrex cards, data sheets, graphs, state-of-the-art reports, reviews, handbooks, or bibliographies. Both personal and group profiles can be negotiated.

SELF-HELPS: The Information Sciences Section provides consulting services dealing with the design and development of personal libraries and information storage and retrieval systems using manual, semiautomated, and computer techniques. A workbook on *Modern Techniques in Chemical Information,* which has chapters on information systems, indexing and classification, secondary sources, reference works, and other useful guidelines, is in preparation. The text should be helpful to metallurgists as well as to chemists.

USER QUALIFICATIONS: None. Most of the services are negotiated on a contract basis.

D16

KNOWLEDGE AVAILABILITY SYSTEMS CENTER (KASC)
University of Pittsburgh
Pittsburgh, Pa. 15213

SPONSOR: University of Pittsburgh, with support from the National Science Foundation and the National Aeronautics and Space Administration. Serves as a NASA Regional Dissemination Center.

PURPOSE AND FUNCTION: A nonprofit computerized information center. In addition to academic instruction and research in the field of information science, KASC provides information services to the University of Pittsburgh, the community, and the nation.

RESOURCES: Searchable magnetic tape data files from a number of sources. On-site are the NASA files (C5), and *CA* Condensates and *Chemical Titles* (F3). KASC also has access, through the NASA Regional Dissemination Center network, to tapes of the Defense Documentation Center (C2), *Government Reports Announcements* (C10), *Engineering Index* COMPENDEX and CITE (F4), *Metals Abstracts* METADEX (B15), and several others of nonmetallurgical interest.

INFORMATION RETRIEVAL SERVICES: Both retrospective and current awareness searches to client specifications.

SPECIAL SERVICES: Workshops to aid in utilization of information; consultation with the University of Pittsburgh School of Engineering; and document service.

USER QUALIFICATIONS: None.

D17

LAWRENCE RADIATION LABORATORY (LRL)
Technical Information Division
Information Research Group
University of California
Bldg. 50B
Berkeley, Calif. 94720
    (See also THERMODYNAMIC PROPERTIES OF METALS AND ALLOYS, D29)

SPONSORS: U.S. Atomic Energy Commission and the University of California.

PURPOSE AND FUNCTION: To provide information resources and literature services to LRL personnel.

INFORMATION RETRIEVAL SERVICES: The computer tapes for *Nuclear Science Abstracts*, from July 1966 to date, have been recorded in the Berkeley Mass Storage System. Under an AEC research contract, the Information Research Group furnishes, twice a month, SDI and on-demand retrospective search services to LRL employees and selected AEC personnel. LRL is one of five AEC sites with a video terminal connected by telephone line to AEC's RECON interactive retrieval system at Oak Ridge, Tennessee. Staff members instruct new users in manipulating the file from the keyboard and also provide written instructions for RECON.

SPECIAL SERVICES: Computer tapes of *Nuclear Science Abstracts* are received at Berkeley one month before the printed issues. From these, the staff prepares computer printouts of author indexes, cumulative report number indexes, and frequency-count listings of index terms and subject categories. The Information Research Group also aids LRL personnel in setting up computerized or other nonconventional indexes for special literature collections.

SELF-HELPS: Microthesauri for various sections of *Nuclear Science Abstracts* are being prepared in conjunction with a project to restructure the *INIS Thesaurus* (see U.S. Atomic Energy Commission, C15).

USER QUALIFICATIONS: Services are free to LRL employees and to selected AEC personnel. Services also are offered to the worldwide nuclear science community on a cost-recovery fee basis.

D18

LIQUID METALS INFORMATION CENTER (LMIC)
Atomics International
P.O. Box 1449
Canoga Park, Calif. 91304

SPONSOR: U.S. Atomic Energy Commission, Division of Reactor Development and Technology.

PURPOSE AND FUNCTION: An information analysis center established to support the integrated national effort known as "Liquid Metal Fast Breeder Reactor Program." Covers properties, chemistry, technology, instrumentation, handling, storage, and safety of liquid metals, with emphasis on sodium and NaK alloys.

RESOURCES: Published material from a wide range of sources is collected, analyzed, interpreted, indexed, and abstracted. This material is then processed for use in a computer-based storage and retrieval system. Reference sources include abstract and index journals, reports, periodicals, bibliographies, literature searches, reviews, and citations, in addition to personal contacts with organizations active in liquid metals work. The collection does not include classified documents or information not available in English.

LIBRARY FACILITIES: The reference collection may be examined on-site by authorized personnel. The staff is available for personal consultation.

PUBLICATIONS: Reports generated by the Liquid Metals Engineering Center (LMEC) and state-of-the-art surveys.

INFORMATION RETRIEVAL SERVICES: Search requests will be answered with computer-produced printouts containing bibliographic data and abstracts.

USER QUALIFICATIONS: Services and publications are available without charge to U.S. Government agencies and their contractors, research and educational institutions, private industry, and members of the general public having a valid need.

D19
MECHANICAL PROPERTIES DATA CENTER (MPDC)
Belfour Stulen, Inc.
13919 W. Bay Shore Dr.
Traverse City, Mich. 49684

SPONSOR: Department of Defense, Defense Supply Agency.

PURPOSE AND FUNCTION: MPDC collects, evaluates, stores, and disseminates structural materials test data for use in materials selection, engineering, design, quality control, analytical programs, and other areas in which detailed data displays are required by the aerospace and defense community. Major emphasis is on metal alloys.

RESOURCES: Detailed data card files, used with preprogrammed computer retrieval and display techniques, contain more than 800,000 test records on 4,000 metals and alloys. Each test record may contain as many as 10 or more data points or values. Between 8,000 and 10,000 test records are added to the system each month. Data sources include published and unpublished technical reports and the available open literature.

PUBLICATIONS:
Aerospace Structural Metals Handbook. Three volumes, updated annually.

Commercial Structural Alloys Handbook (in press).

Inventory Reports (quarterly). A guide to the available mechanical properties data in the MPDC file.

INFORMATION RETRIEVAL SERVICES: Computer-produced data searches display actual test data (such as tensile strength, flexure, creep, and fatigue) and related variables, as well as bibliographies identifying data sources. Output can be in either tabular or graphic form.

SPECIAL SERVICES: Consultation and data analyses for specific applications. Data and storage files may be purchased in either punched card or magnetic tape form.

USER QUALIFICATIONS: Services and publications are available to the defense and industrial complex of the United States. A charge is made to recover a portion of the operating cost.

D20
MECHANIZED INFORMATION CENTER (MIC)
The Ohio State University Libraries
1827 Neil Ave.
Columbus, Ohio 43210

PURPOSE AND FUNCTION: An information services organization designed primarily for faculty and students of Ohio State University; also performs research into information handling methods.

RESOURCES: Searchable magnetic tape files consisting of Chemical Abstracts Service's *Chemical Titles* (F3), *Government Reports Announcements* (C10), Institute for Scientific Information's *Source Index* (E10), and CCM Information Corporation's PANDEX (E3).

INFORMATION RETRIEVAL SERVICES: Both current awareness and retrospective searches. User profiles are entered either by filling out an MIC search request form or by personal or telephone interview. Output is in the form of computer printouts of bibliographic references.

USER QUALIFICATIONS: Primarily for

Ohio State University faculty and students, but services are sometimes provided to outsiders on a limited basis.

D21

NEW ENGLAND RESEARCH APPLICATION CENTER (NERAC)
Mansfield Professional Park
University of Connecticut
Storrs, Conn. 06268

SPONSORS: National Aeronautics and Space Administration and industrial, academic, and public sector clients. Serves as a NASA Regional Dissemination Center.

PURPOSE AND FUNCTION: A central information retrieval capability expert at identifying and retrieving pertinent information from published sources.

RESOURCES: Approximately 20 computerized data bases, either on-site or through tie-in with other specialized information centers. *Metals Abstracts* METADEX (B15), the NASA files (C5), and *Government Reports Announcements* (C10) are on-site. Others of metallurgical interest that can be accessed are: *CA* Condensates and *Chemical Titles* (F3), *Engineering Index* COMPENDEX and CITE (F4), DATRIX (E20), Electronic Properties Information Center (D14), IFI/Plenum (E15), Institute for Scientific Information (E10), *Nuclear Science Abstracts* (C15), Science Information Exchange (C11), and the Defense Documentation Center (C2). Resources also include the major published abstracting and indexing services for manual search.

LIBRARY FACILITIES: NERAC has access to the University of Connecticut library; interlibrary loan and photocopy services.

INFORMATION RETRIEVAL SERVICES: Retrospective and current awareness searches based on interest profiles developed by the client and an experienced staff specialist or consultant.

USER QUALIFICATIONS: None. Services

are provided upon prepayment of an annual retainer fee.

D22

NONDESTRUCTIVE TESTING INFORMATION ANALYSIS CENTER (NTIAC)
Army Materials and Mechanics
Research Center
Watertown, Mass. 02172

SPONSOR: Department of Defense, U.S. Army Materiel Command.

PURPOSE AND FUNCTION: The collection, maintenance, and dissemination of information in the field of nondestructive testing (radiography, ultrasonics, electromagnetics, and related testing methods).

RESOURCES: Acquires and stores the world's published literature on nondestructive testing, plus unpublished reports, memoranda, and miscellaneous documents. These documents are indexed, abstracted, coded, recorded on cards, and organized in a coordinate indexing system.

LIBRARY SERVICES: On-site use of the files and holdings available to qualified requesters. Documents are not loaned.

PUBLICATIONS: *Nondestructive Testing Newsletter* (irregular).
Bibliographies: Report Guides are issued periodically for the various fields of interest. They reflect new and significant publications for a particular field and consist of copies of abstracts, together with pertinent descriptors and accession numbers. Currently, there are 18 titles in the series.

INFORMATION RETRIEVAL SERVICES: Extensive literature searching services as well as lists of literature citations in response to specific requests are available.

SPECIAL SERVICES: Consulting and advisory services; preparation of analyses and evaluations; and furnishing locations of hard-to-find bibliographical materials.

USER QUALIFICATIONS: On-site facilities are open to U.S. Government agencies, research and educational institutions, and industry. Other services are available to those who are entitled to the services of the Defense Documentation Center (C2).

D23

NORTH CAROLINA SCIENCE AND TECHNOLOGY RESEARCH CENTER (STRC)
P.O. Box 12235
Research Triangle Park
Durham, N.C. 27709

SPONSORS: The State of North Carolina and the National Aeronautics and Space Administration. Serves as a NASA Regional Dissemination Center.

PURPOSE AND FUNCTION: An independent state agency that operates a nonprofit regional center for the distribution and dissemination of technical information.

RESOURCES: Searchable magnetic tape data files from a number of sources. On-site are the NASA file (C5), *Government Reports Announcements* and its predecessor, *USGRDR* (C10), and a number of others of nonmetallurgical interest. STRC also has access, through the NASA Regional Dissemination Center network, to *Metals Abstracts* METADEX (B15), *Engineering Index* COMPENDEX and CITE (F4), and *CA* Condensates and *Chemical Titles* (F3).

LIBRARY FACILITIES: Maintains a microfilm collection of NASA reports. Complete files of published abstracts corresponding to the search tapes are maintained and are available for reference. Local libraries provide full-text backup for documents.

INFORMATION RETRIEVAL SERVICES: Retrospective and current awareness searches on client-specified topics; standard interest profiles (regular searches on a specified topic).

USER QUALIFICATIONS: None.

D24

RARE-EARTH INFORMATION CENTER (RIC)
Institute for Atomic Research
Iowa State University
Ames, Iowa 50010

SPONSOR: Iowa State University Institute for Atomic Research, supported by grants from industry. Formerly sponsored by the U.S. Atomic Energy Commission, Division of Technical Information.

PURPOSE AND FUNCTION: To serve the scientific community by collecting, storing, evaluating, and disseminating rare-earth information from various sources. The main interest of the Center is the physical metallurgy and solid-state physics of the rare-earth metals and alloys, although files are maintained on all aspects of metallurgy, chemistry, ceramics, technology, geochemistry, and toxicity of rare-earth elements and compounds.

RESOURCES: The Center has access to more than 12,500 journals and periodicals available at the Iowa State University Library and more than 200,000 U.S. Government reports available at the Ames Laboratory Documents Library. Information from these resources (about 5,000 references in 1970) is stored on punched cards for retrieval.

LIBRARY HOLDINGS: Approximately 3,500 reprints of scientific articles and approximately 50 books.

LIBRARY SERVICES: Resources are available for reference purposes by prior arrangement.

PUBLICATIONS:
RIS News (quarterly).
Special reports, bibliographies, reviews, and compilations are issued irregularly.

INFORMATION RETRIEVAL SERVICES: Most information inquiries are answered at no charge. Extensive surveys or searches and in-depth analyses are performed on a cost-recovery basis.

USER QUALIFICATIONS: Services are available to government agencies, research and educational institutions, and industry throughout the world.

D25

RESEARCH MATERIALS INFORMATION CENTER (RMIC)

Oak Ridge National Laboratory
P.O. Box X
Oak Ridge, Tenn. 37830

SPONSOR: U.S. Atomic Energy Commission, Division of Research; operated by the Solid State Division, Oak Ridge National Laboratory.

PURPOSE AND FUNCTION: Primarily to provide information to materials researchers about the availability, preparation, and properties of high-purity research specimens. In addition to information on preparation and assay, RMIC covers magnetic, electrical, optical, and structural properties which may give clues to structure and purity. Ultrapurification and crystal growth are important topics; engineering and mechanical properties are excluded.

RESOURCES: The collection totals about 46,000 documents (data sheets, abstracts, reports, and papers) contained in 140 100-ft. reels of coded 16-mm microfilm in a Recordak Miracode storage and retrieval system. Each document is indexed by the properties and materials studied. A *visible* file provides a rough and ready listing of the known availability and quality of materials.

PUBLICATIONS: Newsletters; an annual bulletin containing three lists— one of materials available, one of references to crystal growth methods, and one of materials for which no sources were found.

INFORMATION RETRIEVAL SERVICES: Requests for sources often can be filled immediately from the visible file. If available, a detailed data sheet is mailed. For more complex requests, the Miracode file can provide a list of literature references. Both files are available for on-site inspection by qualified researchers.

USER QUALIFICATIONS: Users should be qualified researchers in pure materials. They are urged to contribute to the Center as well as to use its facilities so that it can serve as a clearinghouse or exchange center.

D26

SUPERCONDUCTIVE MATERIALS DATA CENTER (SMDC)
General Electric Research and Development Center
P.O. Box 8
Schenectady, N.Y. 12301

SPONSOR: National Bureau of Standards, Office of Standard Reference Data.

PURPOSE AND FUNCTION: To collect, collate, and disseminate information on superconductive materials and to evolve standard values of important parameters for these materials.

PUBLICATIONS: *Superconductive Materials and Some of Their Properties*, National Bureau of Standards Technical Note No. 482, 1969. Available from U.S. Government Printing Office (C12).

SPECIAL INFORMATION SERVICES: Data tabulated on superconductive materials include critical temperatures, critical magnetic fields, crystallographic descriptions, sources of data, and other information contained in the above publication. These data are recorded on IBM cards for searching. Individual queries are answered, and periodic reports of new data are made available.

D27

TECHNOLOGY APPLICATION CENTER (TAC)
The University of New Mexico
Albuquerque, N.M. 87106

SPONSOR: National Aeronautics and Space Administration. Serves as a NASA Regional Dissemination Center for the Rocky Mountain and southwestern United States.

PURPOSE AND FUNCTION: A nonprofit information dissemination center serving industrial clients and researchers.

RESOURCES: Searchable magnetic tape data files. On-site are the NASA *Scientific and Technical Aerospace Reports* (C5) and *International Aerospace Abstracts* tapes (B10). Through the Regional Dissemination Center network, TAC can also provide searches of other computerized data banks including *Government Reports Announcements* (C10), *Engineering Index* COMPENDEX (F4), and *CA* Condensates and *Chemical Titles* (F3).

LIBRARY FACILITIES: Include a microfilm collection of NASA reports and complete files of published abstracts corresponding to the tapes available for searching. Local libraries, such as that of the University of New Mexico, provide text backup for documents.

INFORMATION RETRIEVAL SERVICES: Offers current awareness and retrospective searches on topics specified by clients; also Standard Interest Profiles.

USER QUALIFICATIONS: None. Cost quotations available on request.

D28

TECHNOLOGY USE STUDIES CENTER (TUSC)
Southeastern State College
Durant, Okla. 74701

SPONSOR: National Aeronautics and Space Administration as a "Special Dissemination Experiment" (NASA Regional Dissemination Center).

PURPOSE AND FUNCTION: A nonprofit center for the distribution and dissemination of technical information and technical assistance to industrial firms and other organizations in a 19-county area of southeastern Oklahoma and a 15-county area in northeast Texas.

RESOURCES: Files of publications including *Scientific and Technical Aerospace Reports* (C5), *International Aerospace Abstracts* (B10), *Government Reports Announcements* and its predecessor, *USGRDR* (C10), and a number of others of nonmetallurgical interest.

LIBRARY FACILITIES: Maintains a microfiche collection of NASA and other government reports; a hardcopy collection of NASA's formal publications, including AGARD (Advisory Group on Aerospace Research and Development) and *Tech Briefs*;

and a number of other publications of nonmetallurgical interest.

INFORMATION RETRIEVAL SERVICES: Retrospective and current awareness searches on client-specified topics.

USER QUALIFICATIONS: Primarily serves the 34-county area in Oklahoma and Texas; the faculties of Southeastern State College, Oklahoma State University, the University of Oklahoma, and other state colleges and universities; and government, public, and private agencies concerned with promoting the economic and technological development of the region.

D29

THERMODYNAMIC PROPERTIES OF METALS AND ALLOYS (TPMA)
Lawrence Radiation Laboratory
Hearst Mining Bldg.
University of California
Berkeley, Calif. 94720
(see also LAWRENCE RADIATION LABORATORY, D17)

SPONSOR: U.S. Atomic Energy Commission, Division of Research.

PURPOSE AND FUNCTION: The Inorganic Materials Research Division of the Lawrence Radiation Laboratory is able to provide limited bibliographies on request, as an offshoot of its research program for AEC. Coverage is limited to metallic systems with emphasis on phase diagrams and crystal structure.

RESOURCES: A file of 7,000 carefully selected references that is used primarily in a project for continuing evaluation of pertinent data. Upon request, a qualified user is permitted to search the files and examine the articles in the collection.

PUBLICATIONS: *Selected Values of Thermodynamic Properties of Metals and Alloys* (Wiley). Published in 1963, with new edition in preparation. Data sheets are issued periodically.

INFORMATION RETRIEVAL SERVICES: Searches are conducted, and bibliographies are provided in the form of

Xerox copies of file cards which often contain abstracts. The field of interest must be specified quite narrowly; requests with broad specifications leading to a large number of references cannot be honored.

USER QUALIFICATIONS: U.S. Government agencies and their contractors, research and educational institutions, and industry.

D30

THERMOPHYSICAL PROPERTIES RESEARCH CENTER (TPRC)
Purdue University
2595 Yeager Rd.
West Lafayette, Ind. 47906

SPONSORS: Department of Defense, Defense Supply Agency; National Science Foundation; National Bureau of Standards; other source funding.

PURPOSE AND FUNCTION: To provide source information on 14 thermophysical properties of more than 60,000 different materials. TPRC estimates that more than 60 percent of its scientific documentation files should be of interest to materials engineers and that more than 80 percent of the 13-volume *TPRC Data Series* should be of interest to metallurgists.

RESOURCES: Approximately 60,000 unclassified technical papers from worldwide sources have been documented, coded in depth, and entered in the tape files. Approximately 5,000 new technical papers are added to the file each year.

LIBRARY FACILITIES: Microfiche reproduction services on approximately 55,000 technical papers stored in the TPRC permanent library are provided for nominal fees.

PUBLICATIONS: *Thermophysical Properties Research Literature Retrieval Guide* (Plenum Press). A 3-volume set providing information for users to make their own literature searches on 13 thermophysical properties. Covers the years 1920 through 1964. Six supplements will cover the years 1964 through 1970.

Thermophysical Properties of Matter—TPRC Data Series (Plenum Press). A 13-volume series of which 9 are published. The entire series is to be completed by 1974 and will contain more than 16,000 pages of graphical and tabular data.

Thermophysical Properties of High-Temperature Solid Materials (Macmillan). A 6-volume series—8,500 pages—covering 12 thermophysical and physical properties of 1,375 solid state material groups, including elements, alloys, oxides, ceramics, and intermetallics.

Masters Theses in the Pure and Applied Sciences Accepted by Colleges and Universities of the United States (annual, 1955– ; University Microfilms, Inc.). Lists titles of masters theses covering 45 disciplines in the physical sciences from approximately 200 major universities.

INFORMATION RETRIEVAL SERVICES: Retrospective literature searches can be made for nominal fees. Descriptive brochures available upon request.

SPECIAL SERVICES: Consulting, experimental investigations, and data predictions will be made upon request.

USER QUALIFICATIONS: No restrictions, subject to payment of normal fees on technical inquiry services.

D31

UNIVERSITY OF CALGARY
Data Centre
Information Systems and Services Division
Calgary, 44, Alb., Canada

PURPOSE AND FUNCTION: Provides the faculty, supporting staff, and students, as well as outside users, with current awareness (SDI) and retrospective search services from both internal and acquired data bases.

RESOURCES: *Engineering Index* COMPENDEX magnetic tape files (F4), ERIC tapes (nonmetallurgical), and some internal data bases.

LIBRARY SERVICES: A small library

serves the staff and students. Other users are referred to the main library.

INFORMATION RETRIEVAL SERVICES: Both Selective Dissemination of Information (SDI)—currently some 200 user interest profiles—and retrospective searches.

SPECIAL SERVICES: The Information Systems and Services Division utilizes several computer search programs, some of which enable it to produce special subject or author indexes, and special abstract printouts for other departments on campus and for outside users.

USER QUALIFICATIONS: None.

D32

UNIVERSITY OF GEORGIA
COMPUTER CENTER
INFORMATION SCIENCE UNIT
Athens, Ga. 30601

PURPOSE AND FUNCTION: To establish an information center primarily to support the research activities of the faculty and students of the University System of Georgia.

RESOURCES: Searchable magnetic tape data files consisting of *Engineering Index* COMPENDEX (F4), *CA* Condensates and *Chemical Titles* (F3), and *Nuclear Science Abstracts* (C15).

INFORMATION RETRIEVAL SERVICES: Retrospective and current awareness searches on interest profiles provided by clients. Output in the form of literature references can be provided on either computer paper or cards.

USER QUALIFICATIONS: None. Somewhat higher fees for commercial clients than for academic users.

D33

WATERVLIET ARSENAL BENET LABORATORIES
TECHNICAL INFORMATION SERVICES OFFICE
Watervliet, N.Y. 12189

SPONSOR: U.S. Army Technical Information Office.

PURPOSE AND FUNCTION: Specializes in material on research and development of the Army's heavy conven-tional weapons, cannon, mortars, and recoilless rifles; the improvement of existing weapons; and the development of new weaponry. Metallurgical coverage includes metallography, brittleness, composite materials, corrosion, crystal structure, deformation, ductility, and elasticity.

LIBRARY HOLDINGS: Approximately 28,000 titles.

LIBRARY SERVICES: Open for reference to visitors with user qualifications shown below; reference and referral services; duplicating; and limited loan services.

PUBLICATIONS: Accession lists for classified and unclassified reports; about 60 technical reports annually.

USER QUALIFICATIONS: U.S. Government agencies and their contractors with need to know and clearance. Academic institutions, other libraries, and the scientific community, in general, may use the library within applicable security regulations and library policies.

D34

WESTERN RESEARCH APPLICATION CENTER (WESRAC)
University of Southern California
809 W. 34th St.
Los Angeles, Calif. 90007

SPONSORS: The University of Southern California and the National Aeronautics and Space Administration. Serves as a NASA Regional Dissemination Center.

PURPOSE AND FUNCTION: A nonprofit technical information center which receives and disseminates technical and scientific knowledge.

RESOURCES: Searchable magnetic tape data files from several sources, including the NASA data bank (*Scientific and Technical Aerospace Reports*, C5, and *International Aerospace Abstracts*, B10), NTIS data bank (*Government Reports Announcements* and *U.S. Government Research and Development Reports*, C10), *Engineering Index* COMPENDEX (F4), and *CA* Condensates (F3).

LIBRARY FACILITIES: A microfilm collection of NASA reports and complete files of published abstracts corresponding to search tapes are maintained. Full-text backup for documents is provided through the University and other libraries.

INFORMATION RETRIEVAL SERVICES: Current awareness monthly reports of customized interest profiles; retrospective searches—custom in-depth searches of specific problems; Standard Interest Profiles—reports on pre-selected subjects of broader interest; and manual searching of all available sources.

USER QUALIFICATIONS: None. Complete details of resources, service delivery times, output format, and subjects of Standard Interest Profiles are available on request.

E

COMMERCIAL ORGANIZATIONS AND PUBLISHERS

E1

ADLER'S FOREIGN BOOKS, INC.
162 Fifth Ave.
New York, N.Y. 10010

PURPOSE AND FUNCTION: A library service for procurement of books published abroad. Catalog No. 196, *Dictionaries, Handbooks, All Languages, All Fields,* contains almost 2,000 dictionary and handbook titles arranged by language and by subject. The subject section on metallurgy lists 18 dictionaries, but only 9 handbooks—5 in English and 4 in German. Their interpretation of "handbooks" appears to be more in the nature of encyclopedias than data compilations such as the *Metals Handbook.*

E2

ASSOCIATED TECHNICAL SERVICES, INC. (ATS)
855 Bloomfield Ave.
Glen Ridge, N.J. 07028

PURPOSE AND FUNCTION: A commercial organization specializing in the translation and dissemination of foreign scientific information in some 35 languages. ATS collects and serves as marketing agent for 2,250 technical dictionaries in 40 languages and currently has available more than 11,000 technical translations. It covers all aspects of metallurgy but is particularly active in nonferrous metals.

PUBLICATIONS:
Dictionary catalogs. Sections on metallurgy, corrosion, welding, and mining include titles of 45 dictionaries.

Lists of Translations (issued irregularly). Covers titles and prices of hundreds of translations issued during the course of a year. There are separate lists for different subject categories, and one list groups translations on "Electrochemistry, Corrosion and Metallurgy—Fuel Cells, Electrode Processes, Passivation, Stress-Corrosion, Radiography, Smelting, Welding, Alloys, Wear."

SPECIAL SERVICES: Translations made to order; current awareness and monitoring services of Russian literature; searches of foreign and domestic literature in connection with patent invalidation, prior art, state of the art, novelty, etc.; and consultation in foreign scientific documentation.

E3

CCM INFORMATION CORPORATION
A Subsidiary of Crowell Collier and MacMillan, Inc.
909 Third Ave.
New York, N.Y. 10022

PURPOSE AND FUNCTION: A commercial organization which produces numerous information products in a variety of disciplines. Also serves as a publishing and marketing agent for the information products of other organizations. Products and services of metallurgical interest are given below.

PUBLICATIONS:

Indexes: World Meetings, U.S. and Canada and *World Meetings, Outside U.S. and Canada* (quarterly). Both provide in-depth information on future meetings of interest to the scientific, medical, and engineering communities. Each issue is completely revised and cumulated and contains listings of future meetings with indexes of date, location, and subject by keyword.

Calls for Papers (weekly). An alerting service to keep the scientist, engineer, and management aware of opportunities for the presentation of oral papers at meetings in the United States and Canada.

Transdex: Guide to U.S. Government JPRS Translations of Iron-Curtain Documents (monthly). Provides a printed guide and index to all documents from Russia, China, Vietnam, Eastern Europe, etc., translated by the U.S. Joint Publications Research Service (JPRS) (see C3). *Transdex* includes bibliographic listings; subject, country, and author indexes; and a reference index for microfilm and microfiche copies of complete translations. These copies are available by separate subscription.

Sci/Tech Quarterly Index to U.S. Government Translations. Indexes scientific and technical materials drawn from books, research reports, newspapers, and journal articles from 125 countries, representing about 60 percent of the translations prepared by JPRS.

SERVICES: PANDEX Current Index to Scientific and Technical Literature (biweekly). A computer-based, microfiche and microfilm service covering annually more than 2,400 scientific, medical, and technical journals; 6,000 books; and 35,000 U.S. Government technical reports. Entries include appropriate material from English, Russian, German, French, and Italian literature. More than 300,000 titles are indexed per year. About 85 percent of the Index is computer-produced from titles, and 15 percent is manually indexed by pre-computer title enrichment. An author index is included. Microfiche versions are issued quarterly and annually. For those organizations with appropriate computer equipment, a magnetic tape edition is available for PANDEX (updated weekly). CCM is also the marketing representative for *Engineering Index's* COMPENDEX tapes (F4) west of the Mississippi (with the exception of Missouri) and in the southern part of the United States, east and west.

SELF-HELPS: The CCM Data Processing Center offers a number of tested computer programs designed for information processing and information retrieval. The Systems Department offers programs designed to create indexes for a customer's own serial holdings.

CCM serves as publisher and sales agent for the *Engineering Index Thesaurus* (F4).

E4

DERWENT PUBLICATIONS LTD.
Rochdale House
128 Theobalds Rd.
London, WC1X8RP, England

PUBLICATIONS: *Central Patents Index*, in 12 *Alerting Bulletins* by subject section, plus a number of patents abstracts bulletins by country.

Central Patents Index Alerting Bulletin M—Metallurgy (weekly). Lists about 7,500 abstracts per year, arranged by country, including patent number, class, alerting patentee, and basic patentee indexes in each issue.

E5

DIFFUSION INFORMATION CENTER
22447 Lake Rd., 205D
Cleveland, Ohio 44116

PUBLICATION: *Diffusion Data* (three or four times yearly). A review publication with chapters covering diffusion in various types of substances and materials including solid and liquid metals and alloys. Bibliographies with each chapter. Author, subject, and formula indexes.

E6

FINISHING PUBLICATIONS LTD.
17 Cranmer Rd.
Hampton Hill, Middlesex, England

PUBLICATIONS:
Metal Finishing Plant and Processes. A primary journal.
Metal Finishing Abstracts (bimonthly, 1959–). Provides about 4,000 abstracts per year from journals, plus patents, books, reports, and translations. Arranged in 20 subject sections, with an author index in each issue and author and subject indexes annually.
Also publishes a number of books.

E7

GALE RESEARCH CO.
Book Tower
Detroit, Mich. 48226

PURPOSE AND FUNCTION: Publishers of reference books, dictionaries, and directories in all fields. Most of the titles listed below were used in compiling material for this book and could prove helpful to the reader who wants supplementary information.

PUBLICATIONS: *Acronyms and Initialisms Dictionary.* Ellen T. Crowley and Robert C. Thomas, eds., 3rd Edition, 1970. A guide to alphabetic designations, contractions, acronyms, initialisms, and similar condensed appellations covering associations, business and trade, domestic and international affairs, government, labor, science, societies, specifications, and other fields. An annual supplement is offered.
Encyclopedia of Associations. Margaret Fisk et al., eds., 6th Edition, 1970, in three volumes. Covers nearly 14,000 national, nonprofit membership organizations. Typical entries in Volume 1 cover 17 vital items including name, address, purpose, and objectives; name and title of chief executive; number of members; publications; and convention schedule. Volume 2 is a 2-part index —geographic and name of association executives. Volume 3 is a quarterly update.
Directory of Special Libraries and Information Centers. Anthony T. Kruzas, ed., 2nd Edition, 1968, in three volumes. Information about more than 13,000 special libraries, information centers, and documentation centers in the United States and Canada arranged alphabetically by name of supporting organization. Volume 2 contains geographical and personnel indexes, and Volume 3 is a quarterly supplement.
Research Centers Directory. 3rd Edition, 1968. A comprehensive guide to more than 4,500 sources of advanced knowledge on campuses and elsewhere throughout the United States and Canada. Arranged by major subject fields; indexed by subject, name of research center, sponsoring institution, and name of research director. A periodical supplement, accompanied by cumulative indexes, is issued.
Standards and Specifications Information Sources. Erasmus J. Struglia, 1965. A guide to the literature and the public and private agencies concerned with technological uniformities. Covers general and government sources of standards information; directories of organizations, agencies, and individuals interested in standardization; societies concerned with standardization; and international aspects of standardization, including organizations and publications. Also contains bibliographies of the literature of standardization, catalogs and indexes of standards and specifications, and guides to periodical indexes that include material on standards. Listings are annotated.

E8

GLOBAL ENGINEERING DOCUMENTATION SERVICES, INC.

3950 Campus Dr.
Newport Beach, Calif. 92660

PURPOSE AND FUNCTION: A commercial supplier of copies of standards and specifications, reportedly with the world's largest library of U.S. military and industry specifications, including a complete file of obsolescent documents dating back to 1946. The collection, from some 85 sources, totals over a quarter million military and industrial documents; hard copy or microfilm can be provided within 24 hours of order.

An alerting service, the *GEDS Weekly Bulletin*, contains the latest information from the DOD *Index of Specifications and Standards*, plus new and revised information from the Departments of the Air Force, Army, and Navy; NASA; U.S. Government Printing Office; and technical societies such as SAE, ANSI, ASME, ASTM, and AWS.

E9

HENRY BRUTCHER TECHNICAL TRANSLATIONS
Box 157
Altadena, Calif. 91001

A multicopy translation agency that specializes in translating articles of interest to metallurgists, predominantly from German and Russian periodicals, although competence is claimed in 15 languages. Nearly 7,000 translations have been produced since 1926. Issues catalogs with subject and author indexes, and bimonthly lists of new translations accompanied by abstract and price.

E10

INSTITUTE FOR SCIENTIFIC INFORMATION (ISI)
325 Chestnut St.
Philadelphia, Pa. 19106

PURPOSE AND FUNCTION: A commercial organization that provides services and publications designed to help researchers cope with their literature and information problems. One of the major precepts of ISI's philosophy is that the various fields of science and technology are interdisciplinary and should not be categorized into arbitrary *a priori* classifications; thus much peripheral but often important material will be included in its metallurgical coverage. ISI's basic resource consists of some 3,800 journals, of which about 70 are of strong metallurgical interest; many others are of peripheral interest.

PUBLICATIONS:

Abstracts, Indexes, and Alerting Publications: Current Contents (weekly). A series of five separate publications serving five major fields of interest. Those of metallurgical interest are *Current Contents—Engineering and Technology* and *Current Contents—Physical and Chemical Sciences*. These publications reproduce photographically the table of contents pages of the most current issues of several hundred journals. A unique feature is an author index (first author only) also listing the address.

Science Citation Index (SCI) (quarterly; fourth issue is annual cumulation). An index of cited papers and their sources. Arranged in two main sequences: *Citation Index* and *Source Index*. The *Citation Index* portion lists, by first author, those papers that are cited by other authors in the list of about 2,400 journals that are covered. Each entry shows cited author, publication year, name of publication, volume, and page number of cited article. It also shows year of citing article, journal, volume number, and page. A typical annual cumulation will provide 4 million index entries to cover 350,000 current items. The *Source Index* gives complete bibliographical citations arranged by author for citing authors listed in the *Citation Index*. Detailed instructions for use of *SCI* are included in each issue. Cumulations are available for 1961, 1964, and annually thereafter.

Permuterm Subject Index (quarterly and annual, 1966–). A multidisciplinary, computer-generated, permuted title word index to the current journal articles covered by *Science Citation Index*.

Current Abstracts of Chemistry and Index Chemicus (weekly). Established in 1960 as *Index Chemicus.* Reports on all significant published papers dealing with new compounds and reactions within 30 to 60 days after their initial publication. Organized by journal, the summaries contain the author's abstract, bibliographic description, molecular formulas, structural diagrams, use profiles, analytical information, and other details. Includes subject, formula, author, and journal indexes. Of peripheral metallurgical interest.

INFORMATION RETRIEVAL SERVICES: ASCA (Automatic Subject Citation Alert). A weekly computerized current awareness service based on the 2,400 journals. Users submit interest profiles. A printout of pertinent citations is provided.

ISI Search Service. Provides retrospective searches of the scientific journal literature for approximately the last 10 years. Output consists of bibliographic references on 3 × 5 cards; payment is on an hourly fee basis.

SELF-HELPS: Magnetic tapes are offered for in-house searching. Two files are available—Source Files and Citation Files—corresponding to the two components of *Science Citation Index.* Annual cumulations with weekly updating are provided. Under special agreements with ISI, those who purchase tapes can sell search services outside their organization.

Original Article Tear Sheet service (OATS). Provides complete copies of original documents from the source journals covered. Copies are in the form of tear sheets taken directly from the journals.

Another way of procuring original documents is by Request-a-Print—a set of postcards that can be ordered from ISI bearing a form request for reprints which the user can mail directly to the author(s) of the paper(s) he wants.

ISI's Who Is Publishing in Science is an annual directory of approximately 160,000 authors included in the ISI coverage. It is indexed alphabetically by author, organization, and geographically, but not by subject.

E11

JOHNSON MATTHEY & CO. LTD.
78 Hatton Garden
London, E.C.1, England
or
608 Fifth Ave.
New York, N.Y. 10020

PURPOSE AND FUNCTION: Refiners and distributors of platinum metals from the Rustenburg Platinum Mines, South Africa.

PUBLICATIONS:
Platinum Metals Review (quarterly, 1957–). Contains technical articles, plus a section on "Abstracts of Current Literature on the Platinum Metals and Their Alloys." About 600 abstracts per year from 450 journals, plus patents.

Also publishes other technical literature.

E12

McGRAW-HILL PUBLICATIONS CO.
330 W. 42nd St.
New York, N.Y. 10036

PURPOSE AND FUNCTION: A commercial publisher of technical journals, books, and reference works.

PUBLICATIONS:
Primary Journals: Some 61 trade journals in a variety of fields, including a half dozen dealing with various aspects of metalworking.

Books: The current catalog contains 155 titles in the category of "Mechanics, Mechanical Engineering, and Metallurgy." These include a number of handbooks, some published for other organizations such as the Society of Manufacturing Engineers (B64) and the American Society of Mechanical Engineers (B14).

Also publishes the 15-volume *McGraw-Hill Encyclopedia of Science and Technology* and the *McGraw-Hill Yearbook of Science and Technology,* the latter designed to supplement and update the *Encyclopedia.*

E13

MSA RESEARCH CORP.
Technical Information Division
Evans City, Pa. 16033

PUBLICATION: *Liquid Metals Technology Abstract Bulletin* (quarterly, 1962–). Approximately 200 to 300 abstracts per year from journals, books, reports, and conference proceedings. No indexes.

E14

PERGAMON PRESS, INC.
Maxwell House
Fairview Park
Elmsford, N.Y. 10523
or
PERGAMON PRESS LTD.
Headington Hill Hall
Oxford, OX3 OBW, England

PURPOSE AND FUNCTION: An international publisher of scientific and technical journals, serials, and books. A major publisher of metallurgical books.

PUBLICATIONS:

Primary Journals: The current catalog contains 22 titles in the category of "Materials Science and Metallurgy." Also publishes a number of cover-to-cover translations of Russian journals including *Physics of Metals and Metallography (Fizika Metallov i Metallovedenie)*.

Books: A special catalog of *Dictionaries, Glossaries, Reference Books* includes multilanguage dictionaries for science and technology.

Abstracts and Reviews: Vacuum (monthly, 1951–). A primary journal which includes a section of "Classified Abstracts." Between 1,500 and 2,000 abstracts per year taken from about 100 journals. Author index in each issue; cumulated author and subject indexes annually.

The Pergamon *Progress in* and *Annual Review* series include a number of titles of metallurgical interest, principal among them being *Progress in Materials Science*, a series of books published irregularly. Volumes 1 through 15 cover 1949–1970. These are comprehensive critical literature reviews on various topics of funda-

mental physical metallurgy. An early volume is *Progress in Metal Physics*. Others in the *Annual Review Series* of subject interest are *Advances in Machine Tool Design and Research* (1963–), *Progress in Ceramic Science* (1961–), *Progress in High Temperature Physics & Chemistry* (1967–), and *Progress in Nuclear Energy, Part V—Metallurgy and Fuels* (1956–).

E15

PLENUM PUBLISHING CORP.
227 W. 17th St.
New York, N.Y. 10011

PURPOSE AND FUNCTION: A commercial publisher of books and serials. Two subsidiaries, Consultants Bureau and IFI/Plenum Data Corp., described below, provide specialized information services.

PUBLICATIONS:

Primary Journals: Plenum's catalog contains 20 technical titles. One of metallurgical interest is *Oxidation of Metals* (quarterly).

Books: The section on "Metallurgical Engineering and Metallurgy" of the current catalog contains 42 titles including the 3-volume *Metals Reference Book*; the sections on "Powder Metallurgy" and "Corrosion Engineering" each contain 8.

SPECIAL SERVICES: The Consultants Bureau is a division of Plenum Press responsible for translations of Russian books and journals. It publishes 88 cover-to-cover translation journals in addition to books and special reports. Journal titles of metallurgical interest are included in the list of prime metallurgical journals on pages 20–25.

Another division, IFI/Plenum Data Corp., 1000 Connecticut Ave., N.W., Washington, D.C. 20036, publishes data collections and provides data subscription services. Among the collections is a 3-volume series of Plenum Press *Handbooks of High Temperature Materials*. The three books are entitled *Thermal Radiative Properties, Properties Index,* and *Materials Index*. Also publishes the

Data Series and other publications for the Thermophysical Properties Research Center (D30).

Data Subscription Services primarily are in the field of chemical patents (only of peripheral metallurgical interest). They include the *Uniterm Index to U.S. Chemical and Chemically Related Patents*, available in printed and magnetic tape versions; microfilm of complete patents contained in the *Index*; and a World Chemical Patent Index, available only on magnetic tape. This division also offers a retrieval service, known as Chemical Patent Service Bureau, for searching patents.

E16

POWDER METALLURGY LTD.
Berk Limited
Canning Rd., Stratford
London, E.15, England

PUBLICATION: *Metal Powder Report* (monthly, 1946–). An abstract journal offering worldwide coverage of journals, books, reports, conference proceedings, and patents. Annual author and subject indexes.

E17

SCIENTIFIC INFORMATION CONSULTANTS LTD.
661 Finchley Rd.
London, N.W.2, England

PURPOSE AND FUNCTION: Specializes in translation of Russian and East European publications.

PUBLICATIONS:
Primary Journal: Russian Metallurgy (bimonthly). Translation of *Izvestiya Akademii Nauk SSSR Metally.*

Abstracts and Indexes: Corrosion Control Abstracts (monthly, 1966–). Approximately 6,000 bibliographical entries and abstracts per year. Translation of the Soviet abstracting journal on corrosion.

Mechanical Sciences Abstracts (bimonthly). Cover-to-cover translation of abstracts in *Mashinovedeniye.* Contains a section on "Strength and

Wear Resistance of Machines and Engineering Material."

Czechoslovak Science and Technology Digest (bimonthly). Tables of contents and abstracts of 30 to 40 journals and irregular publications. Includes the following journals of metallurgical interest: *Hutnicke Listy, Slevarenstvi, Kovove Materialy, Hutnik, Strojirenstvi, Czechoslovak Journal of Physics, Series A and B.*

Design, Engineering Materials & Hydraulic Drives Abstracts. English translation of the annual subject index of approximately 12,000 abstracts published in *Referativnyi Zhurnal.* About 15 percent of the entries are under the headings "Metals, Alloys, Steel, Aluminum."

Index to Forthcoming Russian Books (monthly). Contains selected titles and bibliographical data with about 8,000 entries annually; an extensive section on "Metallurgy, Science of Metals, Physics of Metals."

E18

STECHERT-HAFNER, INC.
31 E. 10th St.
New York, N.Y. 10003

PURPOSE AND FUNCTION: An international bookseller dealing in general, specialized, scientific, and technical materials. One of the catalogs is entitled *The World's Languages—Grammars/Dictionaries.* The section on metallurgy contains 24 titles.

E19

3i COMPANY/INFORMATION INTERSCIENCE INCORPORATED
2101 Walnut St.
Philadelphia, Pa. 19103

PURPOSE AND FUNCTION: A commercial organization engaged in various information science projects and services; provides advisory and consulting services, computer facilities, and commercial service bureau functions.

RESOURCES: Searchable magnetic tape data files including *CA Condensates* and *Chemical Titles* (F3), *Engineering Index* COMPENDEX (F4), and INSPEC *Science Abstracts* (B40).

INFORMATION RETRIEVAL SERVICES: Retrospective and current awareness (SDI) searches. Output takes the form of computer printouts or file cards containing titles, authors, availability, sources, index terms, and—with some files—an abstract. Deliveries can be made weekly, semimonthly, or monthly depending upon the file being searched. 3i Company also has been commissioned to act as the marketing outlet for INSPEC tapes and search services in the United States.

E20

UNIVERSITY MICROFILMS
A Xerox Company
300 N. Zeeb Rd.
Ann Arbor, Mich. 48106

PURPOSE AND FUNCTION: Its principal role is to locate and preserve, in microform, the world's significant knowledge. Its services are designed primarily for librarians, but some are likely to be of interest to individual metallurgists.

OUT-OF-PRINT BOOKS AND BACK FILES OF PERIODICALS: Many scientific and technical publishers deposit copies of such materials with copyright permission to University Microfilms allowing them to sell copies, either in microfilm or Xerox hard copy. The American Society for Metals, for example, makes almost all of its out-of-print books available in this manner, as well as its back issues of periodicals such as *Transactions* and *Metal Progress*.

PUBLICATIONS: *Dissertation Abstracts, Section B. The Sciences and Engineering* (monthly, 1932–). (*Section A* covers the humanities and social sciences.) About 150,000 dissertations annually (in both *Sections A* and *B*) from more than 270 cooperating institutes in the United States and Canada. Arranged in 46 main subject categories, 1 being "Materials Science." A subsection on metallurgy is also included in the main section on "Engineering." Keyword title and author indexes in each issue; annual cumulations; a 10-volume cumulation in preparation.

Copies of complete dissertations can be ordered.

Dissertation Digest (monthly). A listing of bibliographic information only (without abstracts), corresponding to *Dissertation Abstracts*.

American Doctoral Dissertations (annual). A listing of all U.S. and Canadian dissertations, whether included in *Dissertation Abstracts* or not.

INFORMATION RETRIEVAL SERVICE: DATRIX (Direct Access to Reference Information; a Xerox Service) is a computerized search system that compiles bibliographies of dissertations in response to queries based on keywords specified by the subscriber. Request forms and appropriate keyword lists are available.

E21

THE H. W. WILSON COMPANY
950 University Ave.
Bronx, N.Y. 10452

PURPOSE AND FUNCTION: Publishers of indexes and reference works for libraries; not directed toward the individual scientist. Coverage is somewhat superficial in regard to metallurgy.

PUBLICATIONS:

Indexes: Applied Science & Technology Index (monthly except August, 1958–). A cumulative subject index to approximately 225 English-language periodicals in the fields of aeronautics and space science, automation, chemistry, construction, earth science, electricity and electronics, engineering, industrial and mechanical arts, machinery, materials, mathematics, metallurgy, petroleum, physics, telecommunications, transportation, and related subjects. Generous index headings, subheadings, and cross-references. Annual cumulations are available. Preceded by *Industrial Arts Index* (1928–1957).

Bibliographic Index (three times a year, 1937–). A subject list of bibliographies, in both English and foreign languages, that contains 40 or more bibliographic citations. Bibliographies published separately as books and pamphlets are included. In addition, approximately 1,700 periodicals are regularly examined for bibliographic material.

USER QUALIFICATIONS: Indexes are sold to libraries by annual subscription on the "Wilson service basis method of charge," which is governed primarily by the number of publications the library regularly receives.

F

OTHER NONPROFIT ORGANIZATIONS AND MISCELLANEOUS SOURCES

F1

BATTELLE MEMORIAL INSTITUTE
Columbus Laboratories
505 King Ave.
Columbus, Ohio 43201

PURPOSE AND FUNCTION: An endowed institute conducting scientific research on a contract basis for industrial firms and government agencies in almost all areas of science and technology, with extensive programs in all aspects of metallurgy.

LIBRARY HOLDINGS: More than 127,000 volumes.

LIBRARY SERVICES: Open for reference to registered visitors to the Institute; interlibrary loan and photocopy services.

A separate Slavic Library has one of the most comprehensive private collections of Soviet scientific and technical literature in the United States. It is particularly strong in the fields of physics, metallurgy and materials, electronics, and chemistry. The collection dates from 1945 to the present and numbers approximately 12,000 volumes. The Library subscribes to approximately 500 Soviet periodicals, 35 newspapers, 500 East European scientific and technical periodicals, and 175 Soviet journals translated cover-to-cover.

PUBLICATIONS: *Battelle Research Outlook* (quarterly).
Numerous special reports.

SPECIAL INFORMATION SERVICES: Conducts literature searches on a fee basis; provides consultation services; operates the following information centers of metallurgical interest, each described in the directory section noted:

Cobalt Information Center (D7)
Copper Data Center (D8)
Defense Metals and Ceramics Information Center (D11).

Battelle has developed an interactive information storage and retrieval system known as BASIS-70 (Battelle Automated Search Information System). This system has been made available to the Copper Data Center (D8) for its retrieval services.

F2

CHAMBER OF MINES OF SOUTH AFRICA
The Research Organisation
5 Hollard St.
Johannesburg, S.A.

PUBLICATION: *Gold Bulletin* (quarterly, 1968–). Contains technical articles and sections on "Abstracts Selected from Current Technical Literature on Gold" and "New Patents," both worldwide in scope.

F3

CHEMICAL ABSTRACTS SERVICE (CAS)
The Ohio State University
Columbus, Ohio 43210
(See also AMERICAN CHEMICAL SOCIETY, B5)

PURPOSE AND FUNCTION: Stemming from the world's largest abstracting service covering chemistry and chemical engineering, CAS has developed an information system that is modular in concept and characterized by flexibility. CAS is a self-supporting division of the American Chemical Society.

Metallurgical interests are fairly extensive, but must have some relationship to the chemical sciences. Mechanical metallurgy, mechanical properties of metals and alloys, mechanical testing and inspection, heat treating, welding, and mechanical working and forming are largely disregarded. Full coverage is indicated by the 80 subject sections of *Chemical Abstracts* (CA); those of metallurgical interest are noted below.

LIBRARY SERVICES: The CAS library is maintained solely for support of CAS publications and services. However, it can be consulted for reference on a self-service basis. Photocopy service for Russian journals or books.

PUBLICATIONS:

Chemical Abstracts (weekly, 1907–). Currently publishes about 270,000 abstracts annually. Abstracts are informative and appear in 80 subject sections. The information covers more than 13,000 journals, plus books, conference proceedings, patents from 26 countries, and some government reports. Each issue includes Keyword (of subjects), Author, and Numerical Patent Indexes and a Patent Concordance. Volume indexes include variations of these indexes as well as a number of indexes of purely chemical interest. Collective indexes also are available. Index collections were issued every ten years from 1907 through 1956, and every five years since 1956. *CA*

is also offered on 16-mm microfilm.

Chemical Abstracts Section Groupings (biweekly). Five individually published groups of related subject sections from *CA*. Two of these groups, *Applied Chemistry and Chemical Engineering Sections* and *Physical and Analytical Chemistry Sections*, provide most of *CA*'s metallurgical coverage. Of the 18 sections in the first of these 2 groups, 3 are of prime metallurgical interest: Extractive Metallurgy, Ferrous Metals and Alloys, and Nonferrous Metals and Alloys. In the *Physical and Analytical Chemistry Sections*, 8 out of 16 are relevant. The metallurgical interest is more dilute in the following sections: Phase Equilibriums, Chemical Equilibriums, and Solutions; Thermodynamics; Thermochemistry and Thermal Properties; Crystallization and Crystal Structure; Electric Phenomena; Magnetic Phenomena; Electrochemistry; Inorganic Chemicals and Reactions; and Inorganic Analytical Chemistry.

Chemical Titles (CT) (biweekly, 1962–). Reports the titles of selected papers from about 650 chemical journals. Each issue contains approximately 5,000 titles, plus an author index and bibliography. *CT* is computer-produced in a Keyword-in-Context (KWIC) format.

Chemical Abstracts Service Source Index (originally known as *Access*). A directory designed to aid in the location of journals, patents, and meeting proceedings. While useful primarily to librarians, the *Source Index* is a valuable resource of journal identification information, journal holdings data from some 400 libraries, American National Standards Institute title abbreviations, and other useful data. Quarterly supplements are issued.

SPECIAL INFORMATION SERVICES: Many of the CAS publications, or portions thereof, are issued in computer-readable form on magnetic tape for searching. These services are available on lease and are generally accompanied by the corresponding printed journal. *CA* Condensates is the principal computer-readable service of metallurgical interest. The service is issued weekly and provides the following searchable information from the corresponding issues of *Chemical Abstracts*: article titles, author names, bibliographic citations, and phrases from the Keyword Index entries. *Chemical Titles* and the Patent Concordance are also issued in computer-readable form. The latter is *not* available in printed form.

These computer-readable data files are being licensed to a number of the information dissemination centers described in Chapter 5 and directory section D, and CAS encourages individuals to use these centers' services.

SELF-HELPS: In addition to the aforementioned publications, computer-readable data bases are leased to industrial companies as well as to information centers.

F4

ENGINEERING INDEX, INC. (Ei)
United Engineering Center
345 E. 47th St.
New York, N.Y. 10017

PURPOSE AND FUNCTION: Responsible for collecting and maintaining a data base which makes available to all fields of engineering and certain related fields of applied science and management pertinent high-quality research and applications literature. Extensive coverage of the world's literature of metallurgical engineering is provided. However, the areas of more theoretical chemical and physical metallurgy in which no engineering implication is involved are omitted, although the interpretation of "engineering" is quite liberal. A booklet entitled *Ei—The Organization—The Service* is available on request from the Marketing and Business Services Division of Ei.

LIBRARY SERVICES: All publications indexed for Ei products and services are channeled from and to the Engineering Societies Library (A6).

PUBLICATIONS:
Abstracts and Indexes: Engineer-ing Index (monthly, 1962– ; annual, 1887–). Current annual volumes contain approximately 65,000 items; approximately 55,000 are accompanied by abstracts, the remainder by notations. Data are taken from some 3,500 journals, conference proceedings, standards, and monographs. Arrangement is by more than 12,000 main headings and subheadings, liberally cross-referenced. A computer-generated author index appears in each issue of the *Monthly* and is cumulated in the *Annual*. The latter includes a list of publications indexed. The entire file of *Engineering Index* is available on microfilm.

SPECIAL INFORMATION SERVICES: Card-A-Lert is a service whereby all of the entries in *Engineering Index Monthly* are reproduced on 3 × 5 cards and sorted into 167 divisions, 38 "groups" of related divisions, and 6 "disciplines" of related groups. Separate subscriptions can be entered for a division or group, or combinations thereof. The discipline on "Mining, Metals, Petroleum, Fuel Engineering" includes a group of nine divisions on "Metallurgical Engineering—General" and a group of nine divisions on combinations of related metals and alloys.

COMPENDEX (Computerized Engineering Index) (annual, updated monthly). A magnetic tape service, established in 1969, which contains the entire content of the *Monthly Index*, plus additional free-language index terms. (CITE is a related magnetic tape data base containing the input to *Plastics Engineering* for 1965–1969, and *Electrical/Electronics Engineering* for 1968–1969.) These data bases are currently leased or licensed to the following information centers, which can provide searches on request.

3i Company/Information Interscience Incorporated (E19)
Aerospace Research Applications Center (D2)
University of Georgia Computer Center (D32)
University of Calgary (D31).

SELF-HELPS: *Engineering Index Thesaurus.* Created to maintain control of terms used in indexing materials in the electrical/electronics and plastics engineering fields.

Subject Headings for Engineering (SHE) contains approximately 12,000 descriptors under which the abstracts and notations in *Engineering Index* are arranged. This authority list serves as a reference, indexing, and classification tool to aid in the development of literature search profiles; to identify the Card-A-Lert division codes to which subjects are assigned; and to provide an overview of the subject structure of the Ei information services.

Publications Indexed for Engineering (PIE), included in *Engineering Index Annual,* is also available for purchase separately.

COMPENDEX Subsets from the COMPENDEX data base on magnetic tape can be created to match a subscriber's individual profile or specifications. Through this service, corporations may choose specific areas of engineering information for inclusion in their data bases. Organizations and societies may use this service to prepare specialized abstract and/or alerting bulletins.

F5

FRANKLIN INSTITUTE
Science Information Services
20th and Parkway
Philadelphia, Pa. 19103

PURPOSE AND FUNCTION: One of six departments of the Franklin Institute Research Laboratories; established in 1961 to aid industry and scientific institutions in implementing research and development projects. Its main purpose is to provide extensive and accurate literature searches. Maintains a "Metal Sciences Group" that operates in the areas of powder metallurgy, plating and coating, and unconventional machining.

LIBRARY AND LIBRARY SERVICES: An extensive library is maintained to serve all six departments of the Institute. It is open to members of the Institute and to the public, under certain conditions.

PUBLICATIONS:
Abstracts and Bibliographies:
Powder Metallurgy Science & Technology (monthly 1969–). Approximately 2,000 abstracts annually from more than 800 domestic and foreign journals, U.S. and foreign patents, and technical report literature. Arranged in 13 subject categories, with subject, author, and annual indexes.

A monthly *Bulletin* covers unconventional machining developments.

Metallic Pollution Technical Digest (six times a year).

Powder Metallurgy Forging—A Process Evaluation and Bibliography.

Other: NASA Technical Utilization Manual on Powder Metallurgy.

A series of technical bulletins on metal plating and coating, and a series of "information sheets" on powder metallurgy for Pennsylvania industry.

INFORMATION RETRIEVAL AND SPECIAL SERVICES: Literature searches, state-of-the-art surveys, and referral and reference services on specific problems or questions. The New Industries Technical Assistance Program is designed to provide technical assistance to newly founded companies and minority businesses.

USER QUALIFICATIONS: Primarily designed to serve Pennsylvania industries, but has no set geographical limitations.

F6

HUNGARIAN CENTRAL TECHNICAL LIBRARY AND DOCUMENTATION CENTRE
P.O. Box 12
Budapest 8, Hungary

PUBLICATION: *Hungarian Technical Abstracts* (quarterly, 1949– ; in English). Approximately 600 abstracts per year covering domestic journals, reports, and dissertations. Arrangement is by the *Universal Decimal Classification* and includes sections on "Metallurgy, Foundry" and "Properties of Materials." Author index in each issue; subject indexes annually.

F7

INSTITUTE DE SOUDURE
Publications de la Soudure Autogène
32 Bd. de la Chapelle
Paris 18e, France

PUBLICATION: *Bibliographical Bulletin for Welding and Allied Processes* (quarterly, 1949– ; bilingual, English and French). Published for the International Institute of Welding. Approximately 2,500 abstracts per year from about 250 journals, plus papers of the International Institute of Welding. Annual author index.

F8

THE LIBRARY ASSOCIATION
7, Ridgmount St.
London, W.C.1E 7AE, England

PUBLICATION: *British Technology Index*, (monthly, 1962–). Listings are taken from more than 400 British journals; metallurgical interests are well represented. The publication attempts to maintain a high level of currentness. Bibliographic information only is given (no abstract). Articles are listed under alphabetical subject headings, many with detailed subheadings. All keywords in the subheadings are cross-referenced alphabetically. An annual cumulation is provided.

F9

LIGHT METAL EDUCATIONAL FOUNDATION, INC.
c/o Toyo Aluminium K.K.
502 Daiwa Bldg.
Minamikyutaro-machi
Higashi-ku
Osaka, 541, Japan

PUBLICATION: *Metallurgical Abstracts on Light Metals and Alloys* (irregular). Volume V for 1968–69 contains 47 "abstracts" that are actually extended digests in English, many taken from Japanese originals.

F10

THE METAL PROPERTIES COUNCIL (MPC)

United Engineering Center
345 E. 47th St.
New York, N.Y. 10017

SPONSORS: American Society of Mechanical Engineers (ASME), American Society for Metals (ASM), American Society for Testing and Materials (ASTM), and the Engineering Foundation.

PURPOSE AND FUNCTION: To identify major unfulfilled needs for reliable data on the engineering properties of metals and alloys; to plan and conduct programs for collecting, generating, and evaluating such data; and to arrange for making such data available promptly through reports, publications, or other means. MPC, ASME, and ASTM co-sponsor the Joint Committee on Effect of Temperature on the Properties of Metals.

A Technical Advisory Committee plans, approves, and supervises the program, which is implemented by a number of subcommittees. Currently, there are eight subcommittees responsible for projects on Boiler and Pressure Vessel Materials, Quenched and Tempered Steels, Fatigue, Effect of Strain Rate, Relaxation, Irradiation Surveillance Test Program, Fracture Toughness, and Corrosion. There are two Task Groups: one on Effects of Irradiation and one on Ship Propeller Materials. A number of laboratories have contracts with MPC to carry out its projects.

PUBLICATIONS: The work of the Council is made available through publication by one of the sponsoring technical societies. The 1970 Annual Report of the Council lists 16 titles of books, reprints, reports, and data series including sources and prices. Sixteen more titles are in process.

F11

NATIONAL MATERIALS ADVISORY BOARD (NMAB)
National Academy of Sciences—National Academy of Engineering
2101 Constitution Ave., N.W.
Washington, D.C. 20418

PURPOSE AND FUNCTION: Provides a mechanism for defining the materials aspects of national problems and opportunities and making appropriate recommendations for action. One purpose is to provide the means of transmitting materials knowledge and experience across disciplinary and industrial lines. NMAB study projects are carried out by ad hoc committees and panels. There are usually 25 to 40 such groups conducting studies on specific subjects. For example, 80 different studies have been completed on process and equipment problems associated with forming materials such as high-strength steels and superalloys. Other typical study projects include corrosion, composite materials, non-destructive evaluation, theoretical strength of materials, ceramic processing, characterization of materials, high-temperature coatings, fracture prevention, electronic materials, testing for prediction of materials performance, accelerated utilization of new materials, and materials selection procedures.

PUBLICATIONS: Reports on specific materials problems, usually of national scope. Those reports that are unclassified can be obtained from the above address at no charge if copies are available or for a modest charge from either the Defense Documentation Center (C2) or the National Technical Information Service (C10). A list is available of some 36 reports of metallurgical interest published during the past five years.

SPECIAL SERVICES: Staff members are available for free consultation on an informal, limited-time basis.

F12

WELDING RESEARCH COUNCIL (WRC)
345 E. 47th St.
New York, N.Y. 10017

PURPOSE AND FUNCTION: Established by the Engineering Foundation under the sponsorship of the major engineering societies, WRC provides a mechanism for carrying on cooperative research in welding and closely allied fields. The activities of the Council include the administraton of large projects by specific committees as well as sponsorship of small grants-in-aid by the University Research Committee.

PUBLICATIONS:
Welding Research (monthly). Published as a *Supplement* in the *Welding Journal* of the American Welding Society and paginated separately. Many of the articles take the form of comprehensive literature reviews; others give the results of recent research sponsored by the Council and interpretive reports from other sources.

WRC Bulletin (irregular). Contains long research reports and some reports of more limited interest. Eleven titles were issued during the most recent fiscal year bringing the total to approximately 150.

Reports of Progress (monthly). Contains reports of progress on ongoing research, proposed programs of research, translations of important articles, and other pertinent information helpful to research workers.

Welding Research Abroad (monthly). Includes important research findings of the International Institute of Welding and the British Welding Institute, as well as research reports from France, Germany, Russia, Japan, and other countries.

Welding Research News (bimonthly). A newsletter containing condensed, interpretive summations of results of completed and ongoing research in the United States and abroad.

Miscellaneous publications and pamphlets summarizing the results of specific fields of research over a number of years.

USER QUALIFICATIONS: With the exception of *Welding Research*, which is available to anyone who subscribes to *Welding Journal*, WRC publications are available only to WRC subscribers (members), who pay graduated membership fees starting at $175.

F13

ZDA/LDA ABSTRACTING SERVICE
34 Berkeley Sq.
London, W.1, England

PURPOSE AND FUNCTION: Designed to serve lead and zinc users throughout the world. It is jointly financed by the Zinc Development Association and the Lead Development Association of London, the Zinc Institute, Inc., the Lead Industries Association, Inc., and the International Lead Zinc Research Organization of New York. Both abstract journals listed below are sent free of charge to bona fide inquirers. The ZDA/LDA Library contains all items referred to in the abstracts, and a microfilm service is operated to provide copies of most of the original documents.

PUBLICATIONS: *Zinc Abstracts* (monthly, 1943–). Approximately 1,300 abstracts per year from about 350 journals, plus books, standards, reports, and patents; arranged in 22 subject categories with annual subject and author indexes.

Lead Abstracts (monthly, 1958–). Approximately 700 abstracts per year from about 350 journals, plus standards, reports, and patents; arranged in 12 subject categories with annual subject and author indexes.

G

GENERAL REFERENCES

G1

Abstracting Services. Vol. 1—Science, Technology, Medicine, Agriculture. 2nd Edition. The Hague, Netherlands, International Federation for Documentation, 1969. 284 pp. FID Publ. 455.

G2

A Brief Guide to Sources of Scientific and Technical Information. Saul Herner. Washington, D.C., Information Resources Press, 1970. 102 pp.

G3

Directory of the Defense Documentation Center Referral Data Bank. Alexandria, Va., Defense Documentation Center, 1970. 210 pp. AD 712 800. From: National Technical Information Service, Springfield, Va. 22151.

G4

Directory of Federally Supported Information Analysis Centers. Washington, D.C., Committee on Scientific and Technical Information (COSATI), Federal Council for Science and Technology, 1970. 71 pp. PB 189 300. From: National Technical Information Service, Springfield, Va. 22151

G5

A Directory of Information Resources in the United States: Federal Government. Washington, D.C., Library of Congress, National Referral Center for Science and Technology, 1967. 411 pp. From: Superintendent of Documents, U.S. Government Printing Office, Washington, D.C. 20402.

G6

A Directory of Information Resources in the United States: Physical Sciences, Biological Sciences, Engineering. Washington, D.C., Library of Congress, National Referral Center for Science and Technology, 1965. 352 pp. From: Superintendent of Documents, U.S. Government Printing Office, Washington, D.C. 20402.

G7

Directory of Special Libraries and Information Centers. Edited by Anthony T. Kruzas. 2nd Edition. Detroit, Mich., Gale Research Co., 1968. 1,048 pp. (See E7.)

G8

Encyclopedia of Associations. Edited by Margaret Fisk. 6th Edition. Detroit, Mich., Gale Research Co., 1970. (See E7.)

G9

Encyclopedia of Information Sys-

tems and Services. Edited by Anthony T. Kruzas. Ann Arbor, Mich., Edwards Brothers, Inc., 1971. 1,100 pp.

G10

Guide to Literature on Metals and Metallurgical Engineering. Virginia Lee Wilcox. Washington, D.C., American Society for Engineering Education, 1970. 32 pp.

G11

Guide to Metallurgical Information. Eleanor B. Gibson and Elizabeth W. Tapia. 2nd Edition. New York, N.Y., Special Libraries Association, 1965. 222 pp. SLA Bibliography No. 3.

G12

A Guide to a Selection of Computer-Based Science and Technology Reference Services in the U.S.A. Chicago, Ill., American Library Association, Reference Services Division, 1969. 29 pp.

G13

How to Find Out in Iron and Steel. D. White. Elmsford, N.Y., Pergamon Press, 1970. 184 pp.

G14

Industrial Research in Britain. Edited by I.D.L. Ball. 6th Edition. London, England, Harrap Research Publications, 1968. 923 pp. (7th Edition to be published by Francis Hodgson Ltd., P.O. Box 74, Guernsey, Channel Isles.)

G15

Metals Information in Britain. Brian Vickery et al. London, England, Aslib, 1969. OSTI Deposited Report 5035. From: National Lending Library, London, England.

G16

Research Centers Directory. 2nd Edition. Detroit, Mich., Gale Research Co., 1968 (See E7.)

G17

Scientific and Technical Information in Canada. J. P. I. Tyas et al. Special Study No. 8, Science Council of Canada. Ottawa, Ont., Canada, Queen's Printer, 1969, 1970. Part I —Summary and Main Recommendations; Part II—(in seven chapters, available separately) : 1. Government Departments and Agencies, 2. Industry, 3. Universities, 4. International Organizations and Foreign Countries, 5. Techniques and Sources, 6. Libraries, and 7. Economics.

G18

Survey of Scientific-Technical Tape Services. Compiled and edited by Kenneth D. Carroll. New York, N.Y., American Institute of Physics, in conjunction with the American Society for Information Science, 1970. 64 pp. AIP ID 70-3. From: American Institute of Physics, 335 E. 45th St., New York, N.Y. 10017 *or* ASIS Sig/SDI 2, American Society for Information Science, 1140 Connecticut Ave., N.W., Washington, D.C. 20036.

G19

World Guide to Technical Information and Documentation Services (UNESCO). New York, N.Y., Unipub, Inc., 1969. 287 pp.

G20

Worldwide Directory of Mineral Industries Education and Research. Herbert Wohlbier et al. Houston, Texas, Gulf Publishing Co., 1968. 451 pp.

G21

Non-ferrous Metals: A Bibliographical Guide. K. Boodson. London, England, Macdonald & Co. Publishers Ltd., 1972.

INDEX

This is an index to (1) titles of publications, (2) names of organizations providing metallurgical information, (3) metallurgical subjects covered, and (4) types of information resources provided. References to the directory section are indicated alphanumerically; page numbers are given for references to text.

In general, titles of publications and names of organizations are referenced both to text and directory section. Metallurgical subjects are referenced only to the directory section; types of information resources are referenced only to text.

Organizations that cover the broad spectrum of metallurgical topics appear only under "Metals" or "Metallurgy," not under each of the specific topics they encompass. For example, under the term "Aluminum," only those organizations and publications that deal specifically with this metal are listed, although many others (such as American Society for Metals, Institute of Metals, and *Metals Abstracts*) obviously will include important resources dealing with aluminum.

The publications indexed are limited to secondary publications that provide "keys" to information resources (such as abstracts, indexes, reviews, handbooks). Titles of primary journals are not indexed, although many appear in the text (see, for example, the list of prime metallurgical journals on pages 20–25).

INDEX

Forging Design Handbook, p. 28,
B15
Forging Industry Association, p. 8,
B34
Forging Industry Educational and
Research Foundation, B34
Forming, *see* Metal forming
Foundry
American Foundrymen's Society,
B8
*BCIRA Abstracts of Foundry
Literature*, B22
Institute of British Foundrymen,
B37
Non-Ferrous Founders' Society,
B54
Steel Founders' Society of
America, B67
See also Casting
Foundry Sand Handbook, p. 28, B8
Fracture, *see* Mechanical properties
Franklin Institute, pp. 18, 56, F5
Furnaces
Industrial Heating Equipment
Association, B36

Gale Research Co., p. 61, E7
Galvanizing
American Hot Dip Galvanizers
Association, B9
GEDS Weekly Bulletin, p. 38, E8
Global Engineering Documentation
Services, Inc., E8
Gold
U.S. Bureau of Mines, C16
Gold Bulletin, p. 14, F2
Government agencies, p. 41
Government reports, pp. 42, 43
*Government Reports Announce-
ments*, pp. 14, 43, 51, 53, C10
Government Reports Index, pp. 14,
43, C10
*Government Reports Topical An-
nouncements*, p. 57, C10
Gray and Ductile Iron Founders'
Society, pp. 8, 37, B35
Gray iron
American Foundrymen's Society,
B8
*A Guide to a Selection of Computer-
Based Science and Technology
Reference Services in the U.S.A.*,
p. 50, G12
*Guide to Literature on Metals and
Metallurgical Engineering*,
p. 60, G10

Guide to Metallurgical Information,
pp. 20, 27, 60, G11
Guide to Metals Literature, p. 60, A7
Guide to Nuclear Science Abstracts,
C15
A Guide to Personal Indexes, p. 60
*A Guide to Scientific and Technical
Journals in Translation*, p. 33

Hafnium
U.S. Bureau of Mines, C16
Hall Library, *see* Linda Hall Library
Handbook of Chemistry and Physics,
p. 28
Handbook of Electronic Materials,
p. 28, D14
*Handbook of Experimental Stress
Analysis*, p. 28, B63
*Handbook of Industrial Electro-
plating*, p. 28
Handbook of Metal Powders, B50
Handbook of the Metallurgy of Tin,
p. 28
Handbook of Welded Steel Tubing,
p. 28, B69
Handbook of Welding Design, p. 28,
B70
Handbooks, p. 27
*Handbooks of High Temperature
Materials*, E15
Hard facing
American Welding Society, B20
Heating equipment
Industrial Heating Equipment
Association, B36
Heat treating
American Vacuum Society, B19
Industrial Heating Equipment
Association, B36
Metals Handbook, Vol. 2, B15
Metal Treating Institute, B51
National Bureau of Standards,
Metallurgy Division, C6
Heavy metals
U.S. Bureau of Mines, C16
Henry Brutcher Technical Transla-
tions, *see* Brutcher
High-purity metals, *see* Pure metals
High-strength steels
Defense Metals and Ceramics In-
formation Center, D11
The Metal Properties Council, F10
National Materials Advisory
Board, F11
High-temperature materials
*Handbooks of High Temperature
Materials*, E15

169

INDEX

INDEX